D0876410

Carleton College

Given In Memory of
Elizabeth Gorrie
Garmhausen '24

Janet Garmhausen Bock
Betsy Garmhausen Hunter

LAURENCE MCKINLEY GOULD
LIBRARY

FOSSILS ALIVE!

FOSSILS ALIVE!

or

New Walks in an Old Field

A series of ten illustrated time-travel excursions into the geological past, experiencing the faunas and floras, lakes and rivers, geysers and volcanoes that have contributed to the ancient history of Scotland

Nigel H. Trewin

Department of Geology & Petroleum Geology, University of Aberdeen

DUNEDIN

Published by
Dunedin Academic Press Ltd
Hudson House
8 Albany Street
Edinburgh EH1 3QB
Scotland

ISBN 978-1-903765-88-3

© 2008 Nigel H. Trewin

The right of Nigel H. Trewin to be identified as the author of this book has been asserted by him in accordance with sections 77 & 78 of the Copyright, Designs and Patents Act 1988

All rights reserved.
No part of this publication may be reproduced or transmitted in any form or by any means or stored in any retrieval system of any nature without prior written permission, except for fair dealing under the Copyright, Designs and Patents Act 1988 or in accordance with a licence issued by the Copyright Licensing Society in respect of photocopying or reprographic reproduction. Full acknowledgment as to author, publisher and source must be given. Application for permission for any other use of copyright material should be made in writing to the publisher.

BRITISH LIBRARY CATALOGUING IN PUBLICATION DATA
A catalogue record for this book is available from the British Library

Typeset by Makar Publishing Production
Printed and bound in Poland. Produced by Polskabook

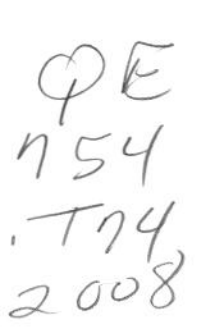
QE 754 .T74 2008

G2

Contents

List of Illustrations

Preface

Scotland has a rich and varied geological heritage, and I doubt whether any country in the world of comparable size can boast such a variety of rocks, and such an extensive record of Earth history. This great richness is one reason why Scotland was at the cutting edge of the emerging science of geology in the eighteenth and nineteenth centuries. Another factor was the extraordinary intellectual talent present in Scotland during that period. Amongst this natural bonanza of geology there are many fossil localities that have yielded exceptionally preserved faunas and spectacular individual specimens, or have provided the material for significant advances in palaeontology.

Scottish highlights in palaeontology must include Lapworth's work at Dobb's Linn near Moffat on the use of graptolites for zonation of the Ordovician and Silurian; the Permian and Triassic 'Elgin Reptiles'; the fossil fish faunas of the Old Red Sandstone and the early Devonian biota of the Rhynie chert from Aberdeenshire. Then there are the eurypterids and primitive fish of Lesmahagow; the first specimen of the conodont animal from Granton and 'Lizzie' the Carboniferous proto-reptile from the Carboniferous of East Kirkton. These are just a few examples of fossil finds that have had an international impact on palaeontology.

This book celebrates some of these examples and adds some more, maybe less famous, but no less interesting examples. I started collecting fossils as a boy in the south of England, and have taught and researched palaeontology at the University of Aberdeen since 1968. The time-travel excursions chosen for this volume reflect my interests in Scottish palaeontology. In some of these personal journeys into the past my interest is research-related, as in the case of the Rhynie biota and the fossil fish of Caithness. Other excursions are inspired by fieldwork, particularly in the company of undergraduate and postgraduate students. Much of the original text was written for relaxation on plane and train journeys over several years.

I have attempted to keep palaeontological detail and technical terminology to a minimum, but it is not possible to avoid all technical terms. There is a language of geology, just as there is for any other science, or even art and drama. Learning a little of any language always increases the enjoyment that can be derived from a subject. A bibliography is provided for the reader who wishes to discover more on any aspect of fossils and ancient environments described in this book. An enjoyable and rewarding experience for the reader would be to visit the fossil displays in the National Museum of Scotland in Edinburgh, the Hunterian Museum in Glasgow and regional museums such as Elgin Museum.

Many people have assisted and inspired me in the completion of this work by their company and discussion in the field and laboratory. It is difficult to single out individuals but it gives particular pleasure to record thanks to Stephen Fayers, Lyall Anderson, Bob Davidson, Jeremy Prosser, Carol Hopkins and Hans Kerp who unwittingly accompanied me on some of these time-travel excursions. Others who have supplied inspiration through discussions of fossils and environments are Euan Clarkson, Dianne Edwards, Hagen Hass, Brian Williams, Gordon Walkden, Malcolm Hole, Ken Glennie and many others, particularly at Palaeontological Association conferences, Aberdeen Geological Society meetings and on fossil-collecting expeditions.

Many thanks to Barry Fulton for assistance with drafting, and to Judith Christie for her patience with my (lack of) computing skills. I am grateful, too, to Adrian Hartley, Ken Glennie, Carol Hopkins, Bill Dalgarno, Neil Clark, Stephen Andrews, Stuart Allison and the British Geological Survey for allowing photographs or specimens to be used as figures. Acknowledgements are given in the appropriate captions. Margie Trewin provided much encouragement, artistic advice on my reconstructions and many useful comments on the manuscript.

Introduction

What was it like? We ask this question about events of last night, a week or year ago, and back into history. Museums, film makers and authors recreate sights or sounds of historic occasions, battles, coronations and everyday life in Scotland. We can visit recreated villages from archaeological and historical periods, and experience for a moment an aspect of mediaeval or Iron Age lifestyle without having to endure the diseases or life expectancy of the time. Generally we wander around manicured excavations of wall bases, and imagine the building plans. Just occasionally surprising detail is revealed, as at the Roman site of Vindolanda on Hadrian's Wall, where remarkable preservation gives us leather, cloth and wood—and even the mundane messages of everyday life: an invitation to a party, a gift of new underpants and a prayer to the gods.

This book takes the reader back much, much further in time, beyond the advent of civilisation, before archaeology and deep into geological time. The fossils and rocks reveal fascinating insights of life and landscape, earthquakes, volcanoes and geysers that graced Scotland millions of years ago. In the imagination the reader must free the fossils from their stony graves, and recreate them at the time they were living plants and animals, parts of ancient Scottish ecosystems. This book takes the form of time travel visits to some famous geological localities in Scotland. Imagine going fishing in a lake 380 million years old, or lying on a sward of primitive land vegetation and watching some of the earliest land-living animals on Earth. Experience a catastrophic earthquake at Helmsdale or track lumbering reptiles in the sand dunes of Elgin by following their footprints in the sand.

I have attempted to keep the technical details of text and illustration as basic as possible. Most of the reconstructions are little more than scientific cartoons, or geophantasmograms, but I hope they provide some food for thought. Fossils chosen for illustration connect the stories with the evidence from the past that we can hold in our hands today.

The preserved evidence on which the excursions are based is inevitably incomplete: a shell, a bone, a ripplemark preserved in sand, specimens jostling in the crowded drawers of museum collections or residing on a mantelpiece, a reminder of a geology excursion or a walk in the country. Finding a fossil excites curiosity and a desire to know more. However, scientific publications in palaeontology can be dull unless there is a professional interest in the subject. The results might be a technical description in a learned journal of a new species of fossil, or an analysis of the trilobites of a group of rocks, maybe the life history of one animal reconstructed from the details of anatomy. This is all good stuff and the essential information on which any grander reconstruction can be based, but it is seldom found in The Scotsman, The Herald or even The Press and Journal.

From the basics of fossil bones we reconstruct skeletons, put flesh on the bones and give an animal habitat, diet and maybe even a sex life. At every stage in the process interpretations have to be made—our facts are the bones; from that point onwards it is interpretation, dependent on 'scientific opinion'. Honest scientific opinion is a subject for debate, or even violent disagreement. Ideas should, and do, evolve. New evidence from bones results in revised interpretation: for example, the thumb spike of the dinosaur *Iguanodon* is no longer interpreted as a horn on his nose.

So knowledge builds on individual animals and on fossil communities. Imagination drives us to reconstruct scenes from the geological past. We put our animals into our best interpretation of the landscape or seascape of the time and depict them behaving in the way we have interpreted from their remains in the rocks. Where then is that magical boundary between interpretation (based on 'facts') and imagination? Most geologists know somebody else who works in their own area of interest and is considered to be 'away with the fairies'. Their interpretations are others' imagination. The outcome is essential debate, argument and revision.

In this book the imagination is allowed to wander beyond interpretation, but not to ignore fact. The reader can treat each excursion through the millions of years of geological time as an exercise of imagination—and leave it at that. Alternatively, the basic evidence on which the story is based can be consulted and the reader can then join the game and modify the stories and experiences. The possibilities are vast, but not limitless. There are always those who will say: 'Ah, but do you know about Y?' or 'I found so-and-so at X.' Such information may change the reader's interpretation, or at least restrict their imagination.

In 1841 Hugh Miller published his book *The Old Red Sandstone*, widely regarded as a classic of nineteenth-century geological literature. His subtitle of 'New Walks in an Old Field' neatly summarises his train of thought: reporting and describing features of fossils and rocks, and then attempting to reconstruct the ancient living animal or environment in his vivid imagination. The stories here attempt to do something of the same, but will certainly lack the benefit of Miller's elegant nineteenth-century prose. However, the sense of geological enquiry and interpretation is little changed from his time to the present day.

This is quite enough philosophy. All the reader needs do is sign up for the excursion and follow the guide. I leave it to them to judge how much is interpretation based on fact, and how much is imagination fuelled by cider or a good malt whisky, and speculative conversations with geologists, zoologists and botanists.

Film makers have milked dry the dinosaur theme for the obvious commercialism of large nasty animals that can be set against human interest and fur bikinis of decreasing size. Hence dinosaur animation has advanced to highly sophisticated levels, and any kid knows it would not be nice to meet a '*Velociraptor*' in Spielberg-mode, but few of the same kids know how small they were. Publicity has resulted in false perception by the majority. We have 'Walked with Dinosaurs' and now the more gullible think they know exactly how *Tyrannosaurus* looked after its young.

Imagine a typical, well-produced television nature film with many excellent film clips, and David Attenborough flawlessly covering the seams to fit well-known footage into a new blockbuster theme. Combine this with scary documentaries on volcanic eruption, spouting geysers, earthquake and hurricane and then combine the two into an epic of miniscule proportions—something maybe near the reality of everyday life at selected localities and time-slots in the geological past of Scotland.

There has to be a David Bellamy element in getting the feet wet and looking closely at life. Nature has to be observed closely, and the reader has to put their eyeball close to the object. The small is just as impressive as the large, it only depends on the distance of the view. Modern museum displays sometimes lose the thread by creating vast areas of open space with a single (maybe superb) object in an impressive and expensive case. People tend to notice such displays from distance and 90% of them never see the glory of the specimen on show. If they are brave enough to go close enough to see the detail, either the guide is in the way or the case is so large that they need 20/20 vision to see the specimen. How about a museum where the observer's head is forced to be thirty centimetres from the objects, and there are plenty of them.

As a child I loved the crowded cases of named invertebrate fossils in The Natural History Museum in London's South Kensington; it was a thrill to find examples of species I had also collected on display. Such displays were good for my interests, but possibly not for the more casual interests of the general public. That part of the museum is now the shop selling fluffy dinosaurs and other tourist memorabilia.

Some of the excursions in *Fossils Alive* contain people to make observations and ask questions. Some of these are my friends, colleagues and co-workers, with whom I have spent many happy days on geological trips, in laboratories, museums and down the pub. I thank them for agreeing to join me on these excursions. Other characters and situations are direct from the imagination. Why not come back in time with us, too? We will visit Scottish locations and experience the natural history and environments of the distant past. Travel will be by time machine, and for the purpose of some stories it is amphibious, can materialise on both land and water, and can be driven as a submersible. On land the vehicle has to be parked, and acts as a base point for excursions. It can hover in Earth orbit and has a number of remote sensing capabilities that are needed to find our localities both geographically and in time. Luckily it is not the Tardis, and is confined to our own planet, and there are no aliens. It takes the place of the traditional geological excursion bus, so we will just call it the Bus.

The ten excursions are arranged in stratigraphic order, starting with the oldest in the early Devonian, and ending in the late Jurassic. A geological timescale (Fig I.1) and a geological map of Scotland (Fig I.2) are provided to help the reader locate the excursions both geographically and within geological time.

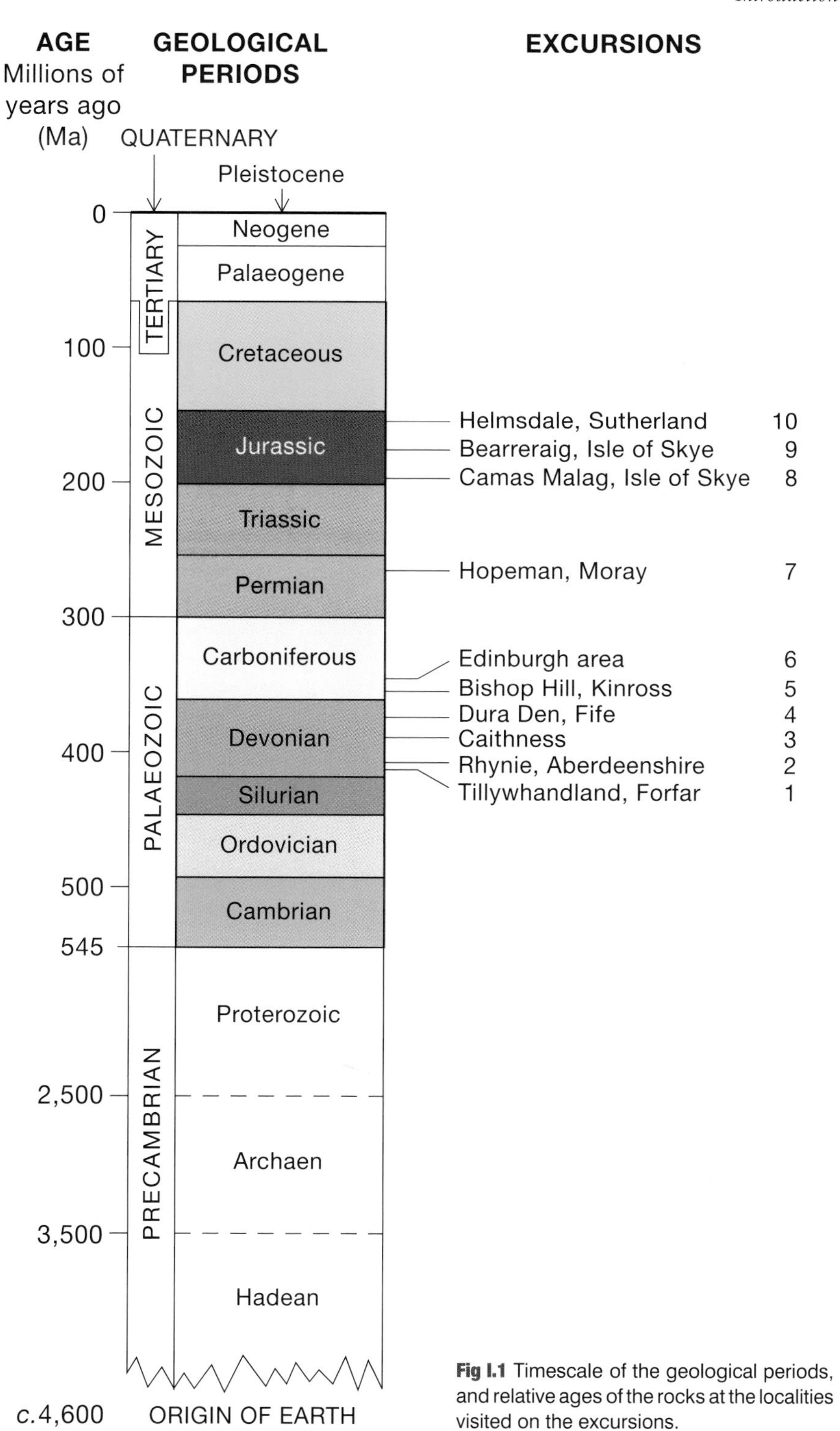

Fig I.1 Timescale of the geological periods, and relative ages of the rocks at the localities visited on the excursions.

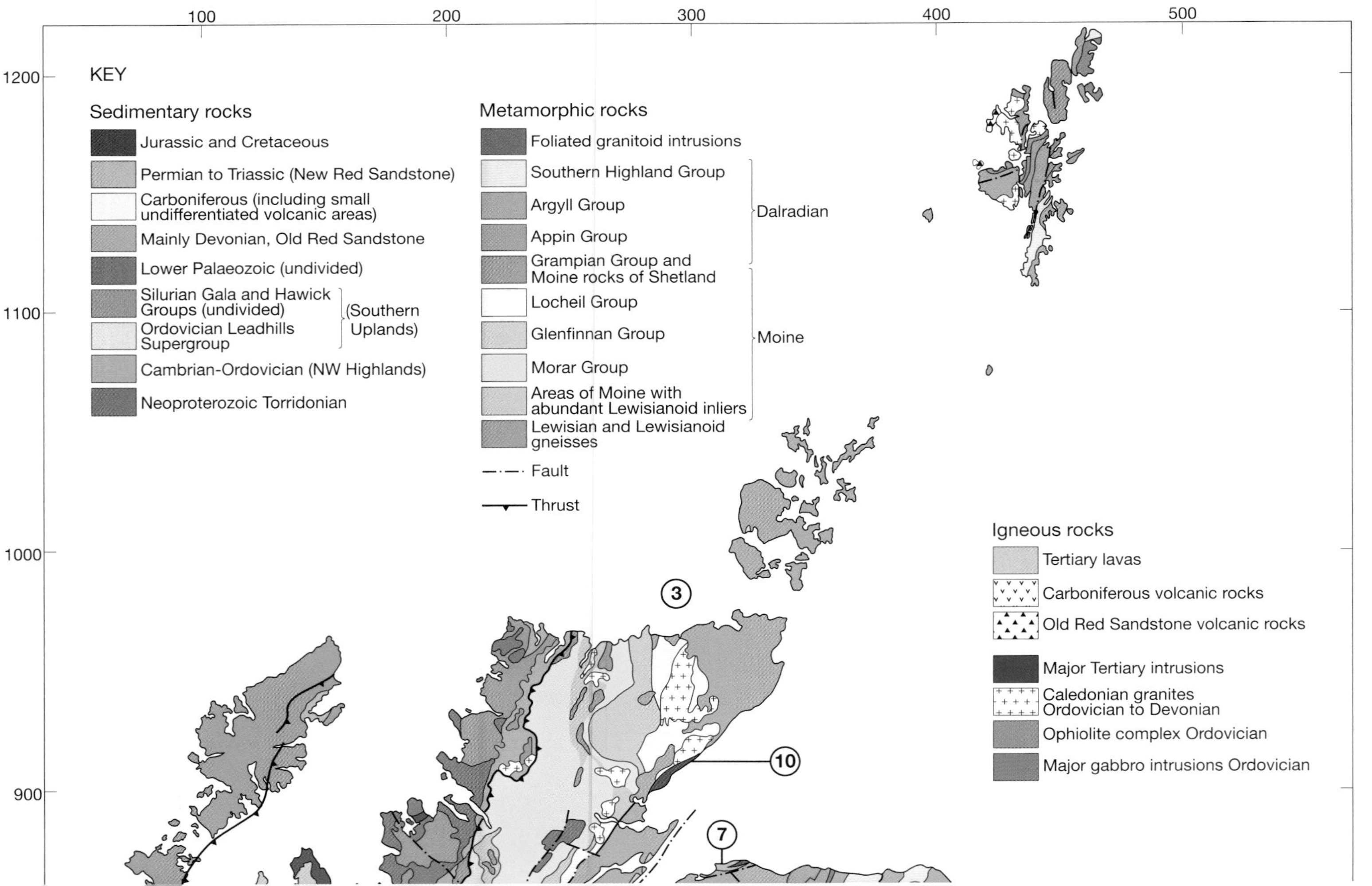

Fig I.2 Geological map of Scotland showing the distribution of rock types and their ages, together with the excursion locations. (Map extensively modified from a British Geological Survey original; coordinates based on National Grid.)

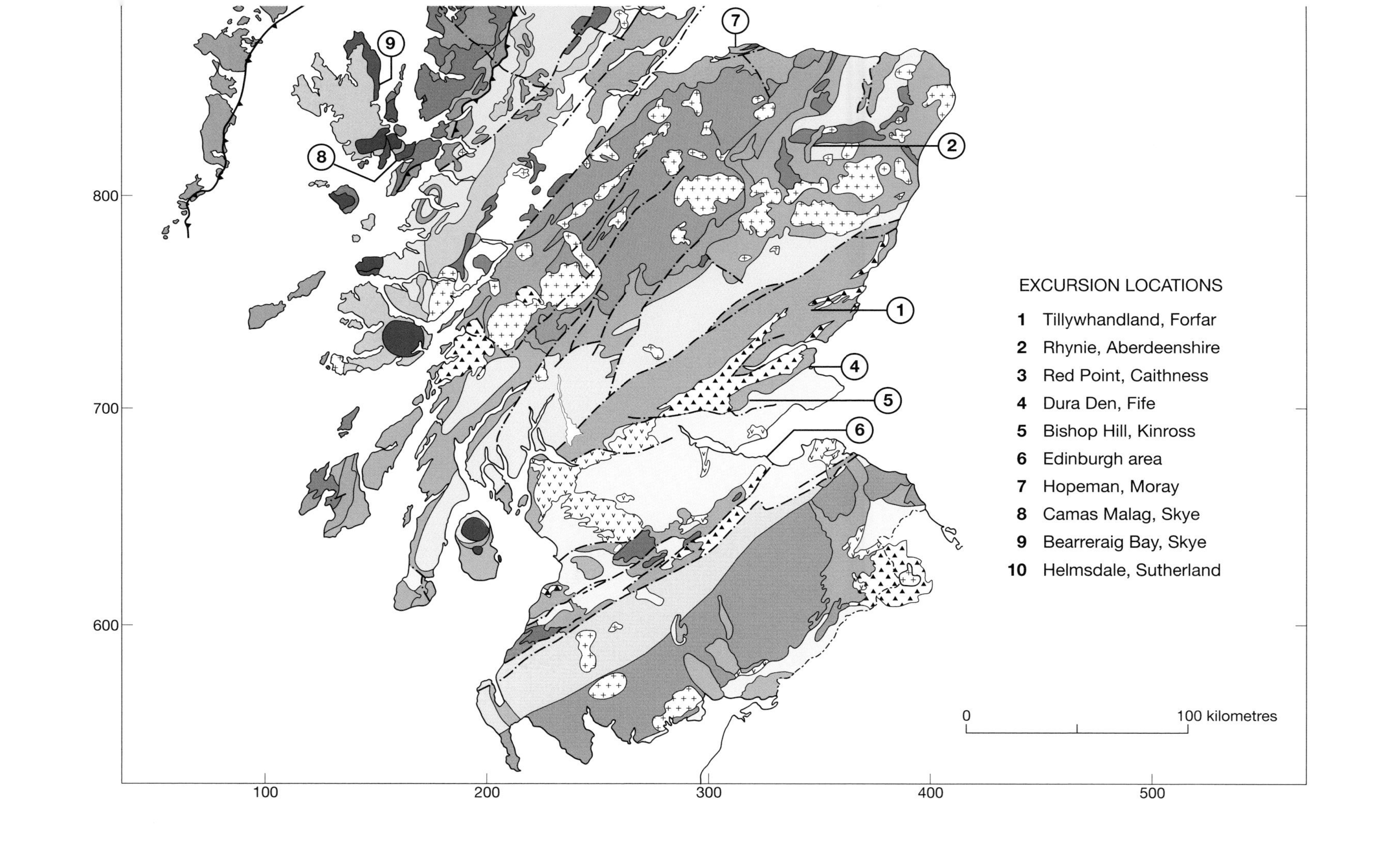
EXCURSION LOCATIONS
1 Tillywhandland, Forfar
2 Rhynie, Aberdeenshire
3 Red Point, Caithness
4 Dura Den, Fife
5 Bishop Hill, Kinross
6 Edinburgh area
7 Hopeman, Moray
8 Camas Malag, Skye
9 Bearreraig Bay, Skye
10 Helmsdale, Sutherland
0
100 kilometres
800
700
600
100
200
300
400
500

Excursion 1

Fish Foray in Forfar

TIME: Early Devonian.

LOCATION: Strathmore, Forfar to Dundee.

OBJECTIVES: To visit the shores of Lake Forfar in the early Devonian and see the fish, arthropods and plants in their natural environment.

THE MODERN EVIDENCE: The deposits of Devonian rivers and lakes and the fossils found in Tillywhandland and other quarries in the Forfar to Dundee region of the Midland Valley.

Prior to about 420 million years ago (Ma) most of Scotland was separated from England and Wales by an ocean. This was the Iapetus Ocean and it had been gradually closing for more than 100 million years. At first the animals in the shallow seas on either sides of the ocean were entirely different, being thousands of kilometres apart, but as the ocean narrowed and closed the faunas became more mixed. This of course was long before the Atlantic Ocean existed; it is only some 60 million years old. The rocks of the northwest Highlands of Scotland have their natural continuations in the Appalachian Mountains of eastern North America, where the fossils are similar to those of Scotland.

The ocean closure and collision of continents was irregular and large bodies of water became isolated during the process as the two continental blocks collided and the intervening ocean was eliminated. This was a time of geological violence. Large faults developed to allow rocks to slide past each other, and earthquakes shook the land. Whilst some areas were rapidly uplifted, others were dragged down to form basins in which great thicknesses of sediment accumulated. The rise of the Caledonian mountains resulted in vigorous erosion that provided vast quantities of mud, sand and gravel to be transported by the youthful rivers which drained the new highlands. Erosion of the mountains was rapid and several kilometres of rock were removed to reveal the granites and metamorphic rocks of the mountain core (Fig I.2). This relic remains as the mountains of Caledonia.

The basins that were cut off by the closure of the Iapetus Ocean initially contained marine life forms such as trilobites and brachiopods that had thrived in the shallow margins of the Iapetus Ocean. Now they were very unhappy. Cut off from the ocean, conditions in the isolated basins deteriorated, the great influx of fresh water draining off the mountains reduced the saltiness of the water, and one by one the marine animals died out and the fauna was greatly reduced in both abundance and diversity.

However, one fauna lost generally results in another gained, and as the basins were converted from cut-off bits of a previous ocean to inland lakes new colonists began to flourish. Maybe they were there all the time and managed to adapt to fresh water; maybe they already inhabited rivers and lakes on the land; or maybe they invaded the new river systems from the marine realm. It is good to speculate, but doubtful whether the evidence still exists to find the truth. However, we have found the fossils of these new colonisers and some were spectacular animals for their time, or any time.

The most interesting newcomers to Scotland were primitive fish and strange arthropods known as eurypterids. They resembled flattened lobsters or scorpions, and some had powerful claws. Fossils of these animals are found in the Lesmahagow area of southern Scotland and are Silurian in age (Fig I.2). On the other side of the Atlantic a eurypterid is the 'State Fossil' of New York, and is similar in age to those found in Scotland.

The relevant observations for our story are those that affected the land surface, revealing areas of uplift and subsidence, marine areas that became isolated from world oceans, and the establishment of drainage patterns on the newly welded continents. Ocean closure and continental collision are not a simple docking manoeuvre: the edges of the continents are irregular, the widths of the shelf seas on the continental margins are variable, and the rocks of the continents themselves are not homogeneous. They are already a mosaic of blocks of differing strength and resistance, frequently bounded by faults, great lines of weakness along which the crust is prone to break. The Great Glen and the Highland Boundary are the positions of such faults that have responded to the ever-changing forces within the crust. Thus the continental collision involved the jostling of blocks the size of the Grampians, and movement, both vertical and lateral, took place along the boundary faults to the blocks.

The reader can try this in two rather than three dimensions in the pub. Make two 'continents' out of dominoes, push them gently together and watch the disruption that takes place as the domino continents are forced together. Domino slides past domino, some rotate from their original orientation. Some may even be forced up and ride over the top of the domino sheets, pushing the experiment into the third dimension. Thus the closing of Iapetus jostled the crustal blocks of Scotland. Where blocks were forced downwards, holes were created that could be filled with the debris eroded from those blocks that were uplifted.

The collision changed the surface drainage radically, and it was not until things had settled down that new major drainage features could become established. Even then there was a lot of geological action, in the form of faulting and also volcanic activity. Continental collision, in which ocean crust is sucked down to great depths, causes melting of rocks; eventually the molten rock rises to form intrusions of granite which do not reach the surface as molten rock but crystallise at a depth of a few kilometres in the crust. Alternatively, molten magma may find its way along weaknesses such as faults, right up to the surface, where a volcano will be formed. Thus, in the aftermath of continental collision, igneous intrusion and volcanic activity flourish. The Cairngorms are an example of a large granite originally intruded into the crust at a depth of about two kilometres, but now exposed at the surface by erosion. The Sidlaw and Ochil hills are largely made of lavas extruded from large volcanoes at the time.

One of the major new river systems established after the closure of the Iapetus Ocean flowed southwestwards down the axis of the present Midland Valley, eventually to find a route to the sea. Marine Devonian rocks occur in Devon and Cornwall so maybe the river found an outlet to the sea some 800 kilometres to the south. The course of the river was controlled by a variety of structural features, channelled by faults along basins between uplifted blocks that formed uplands and mountains. Large volcanoes spewed forth so much material that rivers could be blocked and diverted from their courses.

The object of our journey is the early Devonian environment that existed *c.*415 million years ago in the Forfar–Dundee region of the Midland Valley of Scotland. By this time the large river was well established and flowed southwest. For this time-travel excursion my companion was to be Bob, whose great interest lies in the fossil animal life of the Devonian of Scotland, particularly from the area we were to visit; he is an engineer by training and has worked in the oil industry for many years. Now we found ourselves not going to Tillywhandland Quarry with hammers and chisels to search for fossils (Fig 1.1), but ready to take the Bus to the ancient shore of Loch Forfar for a walk on the wild side of the early Devonian.

Initially we took the Bus back to *c.*425 million years ago in the Silurian and surveyed the scene from a high stationary orbit above the Grampian area. To the far south remnants of the Iapetus Ocean were visible, and large bodies of water

Fig 1.1 A party of vertebrate palaeontologists on a conference excursion searching for fossil fish, arthropods and plants on the tips at Tillywhandland Quarry near Forfar. This locality has produced many excellent early Devonian fish, mainly in the nineteenth century when the quarry was being worked.

that were now isolated from the marine area glinted in the sun. To the north young mountains were being eroded and the resulting sediment transported south by rivers to fill the areas cut off from the ocean. Hilly areas with sharp boundaries seemed to be controlled by active faulting, and disrupted the river drainage of the area, and there were also areas with large conical volcanoes. We needed to adjust our position in time to *c.*415 million years ago, and find a large lake in the present-day Forfar–Dundee area. Then we could land the Bus and examine the ancient environment that we had studied through the rocks of the Lower Old Red Sandstone exposed in quarries of the area.

We adjusted the time factor and positioned the Bus over Forfar. The area was one of lakes and rivers with hilly country to the northwest. The most obvious feature of our new view was a large volcano in the region of Montrose, a volcano that had poured out lava and volcanic ash hundreds of metres thick, and had deeply eroded gullies down its flanks. Further south in the Sidlaw and Ochil

Fig 1.2 Sketch of Lake Forfar area in the early Devonian. The lake is fed by a river carrying sand and mud. A large volcano erupts producing a towering column of ash; collapse of the column results in an intensely hot pyroclastic flow that rushes down the volcano, burning and burying everything in its path.

Hills other large volcanoes were active. Our river snaked its way as best it could around the volcanoes and past uplifted fault blocks. In some places there were large lakes, and one lay below us; this must be Lake Forfar of the early Devonian. The river divided into the distributaries of a delta where it entered the lake (Fig 1.2). As the river slowed it had clearly dumped vast quantities of mud and sand, now represented by the sandstones and shales of the Lower Old Red Sandstone of Strathmore.

We aimed to explore the lakeshore and river banks as far as the environment permitted. There was a lot of water about and we would not be equipped to wade or swim muddy rivers. Bob had spent hundreds of hours fossil collecting in the quarries of the Forfar area, and the prospect of seeing the animals and plants he had collected as fossils, frequently only in a fragmentary state, was a long-cherished dream. Our observations could become a source of personal embarrassment since we had published a brief interpretation of the environment back in 1996, and we could easily find the scene was not as we had interpreted. Never mind, we had used the evidence available at the time, and all science moves on; new knowledge frequently results in new interpretations. Only creationists can ignore the facts, stick to their blind faith and point out that expert scientists change their minds. It always surprises me that creationists do not marvel at the new truths constantly being revealed to mankind by the Maker. Blind faith is easy; reasoning is always more difficult.

Our excursion was about to begin. Here we were in the Bus, watching the surface view of early Devonian Lake Forfar and trying to pick out a possible landing site. This was not easy. A few thousand years' difference in time showed great changes in the area of water in lakes, and the general coloration of the landscape. There seemed to be an alternation between wetter and drier periods, and we certainly wanted to be there when the lakes were present and the rivers flowed. The Montrose volcano was also clearly active, with large-scale eruptions of ash and lava; hopefully we could land during a quiet period. Another problem was just that of distance—how far we would have to walk. We hoped to find lake-shore and river features close enough together to examine both, without being trapped on an island in the lake and river system. Eventually we settled on an area of yellow (and we hoped firm) sand that would provide a safe landing, close to a small river channel running into the lake. We should be able to see something of the lake margin, look at the exposed land surface and peer into the streams.

Our landing was very soft, almost too soft, and for a second there was the horror of landing in quicksand or mud, and sticking—not a nice prospect to be stuck in the mud of the early Devonian. Cautiously we emerged into the Devonian day, the Bus was fine, only sunk in about ten centimetres. It was just a very good landing. We were on sand, sand deposited by a recent flood of the river. Ripple marks covered the surface on which we stood, and in the hollows between the ripples lay broken pieces of plant stem, neatly aligned along the troughs of the ripples. Some fragments were brown and withered, but others still showed green, another indication of the recent flood.

'What if another flood comes?' enquired Bob.

'Not worth considering; it could wash us and the Bus away. The floodwater is very shallow here, so let's go to the channel first and check the level there. We should be able to tell whether the water is rising or falling at present.'

Although our landing area had been flat, the only topography was that of the river channels and sand bars, and whatever depth of water was present in the lake which we could see downstream of our river channel. It was just as well that our landing site had been chosen close to the lake edge and at the margin, rather than within the area of river channels through which water flowed into the lake. These channels were clearly too deep for us to cross, despite the fact that we knew that there were no sunken trees and roots to trip us up, and no large boulders.

A brisk, 200-metre walk to our river channel was the first order of the day. The ripples continued until we were at the slope into the river channel, where they increased in size to small dunes about thirty centimetres high and one metre long. The channel was wide, about fifty metres, with sand bars emerging from the water. A series of parallel terraces around the bars and along the banks chronicled the recent falls in water level following the flood. The river would need to rise nearly two metres to flood the landing site. We felt confident that we would not be flooded in the next few hours.

With the water looking only slightly muddy it did appear as though the flood was past and we were visiting at a time of low flow in the channel. At least we had already seen that the river was prone to flooding, and large quantities of sand were deposited by the floods. The plant debris had been well sorted by the flood and most of the fragments were not identifiable, just the plant hash that is so common in all rivers and typical of the plant debris to be found in many quarries in the Forfar area (Fig 1.3). However, here and there were distinctive fragments showing the characteristic branching and spore cases of *Zosterophyllum*, the tiny branched Y-shaped stem of *Cooksonia* and flat, rather leathery discs of *Parka* covered in round spots (Fig 1.4). It was tempting to sift through this transported debris, examining it with a lens for tiny details that had puzzled palaeobotanists for a century, but we needed to move on and find the living versions in their natural habitats.

The channel we were looking at was the most marginal of many in the channel system, and by following the bank of the channel we could reach the lake without needing to wade rivers. The initial impression was a scene of mud, sand and more mud. There was a lot of sediment exposed on the surface, yet it was generally a damp environment. Browns, yellows and greys were the dominant colours of the landscape with distinct hills to the northwest appearing dark and rocky. To the southwest loomed a great volcanic edifice, the Montrose volcano. It was enormous, with gullies extending down from near the summit. A wisp of smoke, or steam, rose from the summit and from a prominent fissure on the flank. Dark lava fields extended from the foot of this volcanic mountain.

'It looks a bit fresh and active,' commented Bob.

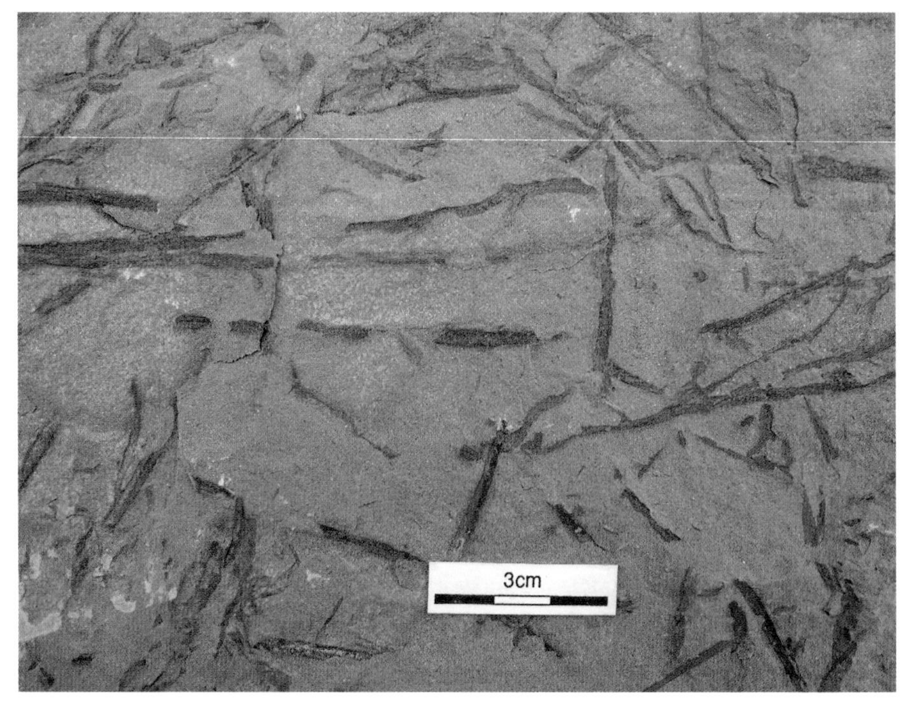

Fig 1.3 Plant fragments in shale from Tillywhandland Quarry near Forfar.

'Yes, but the chances of us picking an eruption must be pretty slim.'

'Maybe, but somebody wins the lottery most weeks.'

Ignoring the possibilities we diverted our attention from the volcano to patches of bright green that bordered the lake—clearly areas with growing plants. The plant fossils from the Forfar area are seldom easy to interpret because most are preserved as small transported fragments, mainly of fairly nondescript plant stems which defy identification. The reason for this is that in order to identify the primitive plants of the early Devonian a specimen is needed which shows the manner of branching of the stems and the type and position of the sporangia—the structures that contained the spores by which these plants reproduced. So the material we saw concentrated in ripple troughs near the Bus was typical of the fossils that are found in the Forfar area, but the patch of fresh green at the water's edge we were about to examine contained the growing plants. There were bound to be surprises, even for us with our limited botanical skills.

We were soon cautiously approaching the lake margin. The ground was distinctly soft in places and mud clung to our boots, curling up the sides and clarting the soles. The plants clearly liked water and at first sight represented a miniature reed bed about thirty centimetres high. Closer examination showed that

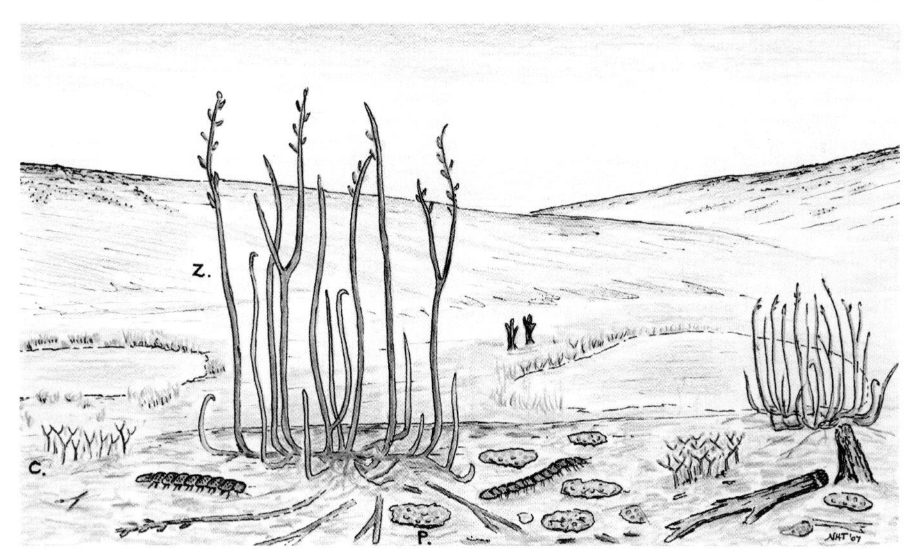

Fig 1.4 Sketch of life on land in the early Devonian of the Forfar area with the plants *Parka* (P), *Zosterophyllum* (Z) and *Cooksonia* (C), and early terrestrial millepedes.

the plants were green naked stems which arose from a branching matted tangle of stems, so closely packed that it was impossible to identify individual plants. The plants spread their net of creeping axes, from which arose the vertical stems. Some stems bore kidney-shaped capsules (the sporangia) in various stages of development. Some were green and immature; but on the older stems they had turned brown and split open, so that as we touched them a delicate cloud of spores was released on the breeze. This was a mini-forest of *Zosterophyllum*, probably the most common plant fossil found at Tillywhandland, but seldom well preserved. At the margins of the marshy areas with *Zosterophyllum* the vegetation was even smaller and close examination was required.

Down on hands and knees we could see the delicate divided stems of *Cooksonia*, each with a knob on the end representing the sporangia. They were very soft and it is not surprising that they are rare as fossils in the area. Between the *Cooksonia* the ground surface was covered by greenish irregular discs with crinkled edges; in some places they were so closely packed that the edges curled up, looking like nondescript American salad. The surfaces of the discs were covered in spots, closely spaced and ranging from green to brown. In the brownish ones the spots were depressions, but in the green ones they appeared as round, disc-like pods. Cutting open the discs revealed a mass of unripe spores. These were clearly sporophyte plants but lacked a stem and formed a horizontal sheet covering the bare surface. They were thus prone to being ripped up in floods and carried with sand

and mud at the will of the current. Hence we now find this curious plant, known as *Parka*, mixed with *Zosterophyllum* fragments in sandstone deposited by floods. We also find *Parka* in fine shales where the plants drifted on the surface of the lake before becoming waterlogged and sinking down through the lake waters to rest in the muds of the lake floor.

Here and there larger plants grew above the general level of *Zosterophyllum*. They had scrawny stems several centimetres in diameter, and branched and tapered rapidly. They were green but through a hand lens they seemed to have a fibrous structure throughout.

'What are they?'

'Goodness knows. It would be nice to be able to take one back to examine, but we can't. I guess it's an alga or lichen or something similar, maybe related to the nematophytes at Rhynie. The smell is rather musty, so it could even be a fungus.'

'Hey, look here,' interrupted Bob, 'there's something much more interesting.'

He was inspecting a sandy hollow that had filled with water in the recent flood; now it was nearly dry and mud cracks were starting to form. Lying on the surface was a crescentic object with a distinctly fishy tail.

'A real cephalaspid, and not long dead!'

'That's *Zenaspis pagei*,' announced Bob after a brief examination. 'He must have been stranded here in the flood and failed to escape.'

'Now it's drying out nicely—welded into the mud on which it perished. All it needs is a covering of mud and sand from the next flood and you have that great fossil rarity—a complete cephalaspid. There are very few complete specimens of these fish in museums.'

Having invested so many hours for science at Tillywhandland Quarry before making this excursion, we had found fossil fragments of this primitive jawless fish, usually only the crescentic headshield with two closely spaced eyes staring upwards (Fig 1.5). The body of the fish was seldom preserved, and probably became detached from the head as a dead carcass drifted in the lake or was swept along by the river. Indeed, the specimen at which we were gazing could suffer the same fate if the next flood eroded the floor of this hollow and swept away our potential fossil. The chances of preservation were certainly slim.

Nevertheless, some magnificent fossil specimens of this type of fish are known, with several specimens closely packed together and still entire. This mode of preservation probably results from the fish being suffocated by the sediment carried in a flood and virtually buried alive as they struggled in mud and silt

Fig 1.5 Headshield of the cephalaspid fish *Zenaspis*, found in Lower Old Red Sandstone, Balruddery, near Dundee.

dumped rapidly by the floodwaters. Finds of such events preserved in the geological record are extremely rare.

Neither of us wanted to desert our poor stranded *Zenaspis*, but we certainly hoped that quarrying, or a new road cut, would one day reveal our friend in the twenty-first century. We could muse on our exact location, but even if we could narrow that down it was most unlikely that our friend would be close to the surface in our lifetime. He could have been eroded thousands or millions of years before Scotland was inhabited, or he might be deeply buried 300 metres under a vast, boring field of barley in the rich farmland of Strathmore. Enough musing, wishing and hoping: there must be more to see than one deceased cephalaspid fish.

We traversed numerous similar hollows with mud-cracked floors on our route beside the lake, but saw no more stranded fish. Clearly we had been lucky, and the fish unlucky. Plant debris was common, and Bob picked up a sharp fin spine from an acanthodian fish. He pronounced it to be the first dorsal spine of an *Ischnacanthus*—I was happy to take his word for such details.

Approaching the lake the sand became damper and the mud was still wet in the hollows, and we trod cautiously as the surface softened. The lake shore was seemingly unremarkable. We were on a small sand spit built into the lake, and ripples on the blue lake waters lapped gently on the sandy shore. The water here was reasonably clear and the bottom was visible to a depth of about a metre. Farther out the water was clearly brown and sediment-laden where the subsiding floodwaters flowed into the lake from the river channel.

The strandline on the shore seemed pretty dull at first sight: a line of plant debris with the usual bits of *Zosterophyllum* and spotted discs of *Parka*. Such debris was concentrated in small bays, and in one of these we stopped to sift through it in the hope of more interesting finds. We were not to be disappointed. Amongst the plant hash was a distinctive crescentic curl with a shiny brown lustre—a millipede. It lay there amongst the plants, standing out by virtue of shape, texture and colour. Having found millipedes before in the Old Red Sandstone, our brains were programmed to notice the familiar, and it was helpful to find our customary fossils as we had hoped and expected. However, the brain is good at well-known shapes, but not so useful for the unfamiliar. If someone does not know, or has never seen, what they are looking for, it is more difficult to track down. Time and again it is the expert in a particular group of fossils who is able to find them in the field—detailed familiarity with the shapes of all the components of the animal aids their recognition. The brain and eye combine to recognise expected patterns, so the finding of something new by casual observation generally requires that it be large and obvious or at least distinctive.

Maybe we recognise millipedes from the Devonian because they are virtually the same as the ones that inhabit dark damp corners in our gardens today. Our dead millipede lying on the Devonian lake shore probably drifted to the shore, having been swept into the lake with plant debris from the flood. It was unlucky to be dead, since millipedes can float on their sides for many days before drowning. They 'breathe' air through rows of holes, 'spiracles', on each side of the body, so when floating on their side one set of spiracles can still function. Maybe the lake surface had been too rough and the poor beast had been swamped.

We continued our search, and Bob made the next discovery. Lying on the sand was a massive claw, as large as the big claw on a good lobster, but rather more delicate. It had sharp opposed points and many serrations between the point and the hinge. Bob picked it up and gently articulated the claw. The sharp serrations sheared past each other; it was an effective grasping and cutting weapon and instantly recognisable as the claw of *Pterygotus*, a massive eurypterid which grew

Fig 1.6 Reconstruction of life in the early Devonian waters of Lake Forfar at the time of formation of the fish bed at Tillywhandland Quarry. Two large *Ischnacanthus* hunt a shoal of *Mesacanthus*. To the left a bottom-dwelling *Zenaspis* swims a safe distance from the large *Pterygotus* seen in the foreground. At the right a *Zenaspis* headshield lies half buried in the mud, while drifted and waterlogged plants (*Parka* and *Zosterophyllum*) lie on the lake floor.

to more than 1.5 metres long (Fig 1.6). It was probably the largest animal about at the time—the early Devonian equivalent of the Cretaceous *Tyrannosaurus*—but not with such good potential for film makers or animation documentaries.

'What's that?'

'Where?'

'In the water—that black shadow—it's moving.'

'Yes, I see—it's cruising slowly along the shore.'

'I agree, yet I can't see any form to it.'

We were looking into the sun, and the ripples reflected the sunlight breaking up the image. The formless black shadow continued to cruise, but suddenly there was a disturbance to the ripples beyond the shadow and a flat back momentarily appeared in the disturbance.

'Of course! We are watching a shadow, a shadow cast by a live *Pterygotus* on the lake bed. The animal is virtually invisible to us due to surface refraction and its excellent camouflage. I have seen the same effect when trout fishing—you cast a fly towards a rising trout, and you see its shadow approach the fly, but you do not see the fish until it breaks surface to take the fly.'

It was difficult to follow the giant arthropod or see how it was swimming—the reflections on the water confused the view. At times it seemed virtually motionless in the water—was it just being a lazy swimmer or lurking with intent? This was frustrating, so near to, yet so far from clear observation of the beast.

'Fish!' We announced in unison. Telltale flashes in the water, and a couple of small fish breaking the surface. There had been 'interaction' between our beast and the fish. Had it caught one in its pincers by ambush? Our answer was 'probably' since the beast sank to the bottom and covered its shadow. We could now see more clearly, and the pincers were clearly working at something.

'I think it has just grabbed a *Mesacanthus* and now it's ripping it up and chewing the bits in its jaws.'

'Hmm—maybe.'

We were just about to give up watching our rather shy subject when it suddenly moved off. The broad fluke of the abdomen gave a flick and it was quickly out of our sight. The swirl of water left behind disturbed the fine sand, but there was something else—shiny fragments glinted in the water. The swirl brought the bits close enough to the shore for us to catch one, and there was our evidence: a ripped piece of acanthodian skin, with the tiny scales giving an iridescent refraction of the sunlight. The predator had clearly had a snack.

There has been much speculation regarding the habits of *Pterygotus*. Some have interpreted it as a bottom-dwelling animal, lurking on a lake bed or in the slack water of a river pool, and grabbing prey by ambush. Its flat shape allows this large animal to present as little area to the current above it as possible. An alternative hypothesis is that the *Pterygotus* cruised gently in mid-water or close to the surface, maybe actively stalking its prey, or just grabbing anything that came within range of its fearsome pincers. Like many such interpretations, both may have an element of truth. The complete specimens of *Pterygotus* that have been found occur in sandstones deposited in rivers; they are also associated with mud flakes—usually a sign of flood conditions. There is one classic specimen in The Natural History Museum in London with two animals preserved side by side. It seems possible that in seeking shelter in a hollow from sediment-laden floodwaters they were suffocated and overwhelmed. Alternatively, they may have been seeking shelter outside the river channel, in shallow, quieter, water and have become stranded as the floodwaters fell—trapped in the same manner as the cephalaspid fish that had already excited our interest.

We had indeed been fortunate to see the eurypterid catch and consume a small acanthodian, and the event provided another scrap of information relevant to the fossils Bob had found at Tillywhandland Quarry. Whilst one always hopes to find a complete specimen of a fossil fish, more often than not the fossil is fragmentary.

In the case of the small acanthodian fish it is common to find small isolated patches of scales. These are hardly worth collecting as fish specimens, but are of interest from the viewpoint of the palaeoecology of the rocks containing the fish. It is possible that these scraps of fish skin and scales represent the remnants of the meals of eurypterids, analagous to the shiny debris left by our large flat friend following his small fish supper.

Time was slipping by and we were more than two hours into our Devonian excursion, and there had to be more to see. We walked for some time along the lake at a gentle stroll, closely examining the material cast up on the shore, but it was generally a matter of the same observations as before. There were a few plant fragments we could not recognise, but all very fragmentary, and only the ability to take samples back and make detailed microscope preparations would aid our study—and this was not permitted. We gained the impression that the plant life was abundant, but that diversity was rather low.

We did see things that are not preserved in the rocks, such as green filamentous bacterial slime on wet surfaces, and similar green and brown mats in stagnant pools—rather like those that might infest a garden pond in summer, something similar to blanket weed.

Such life is not pretty, but lies close to the base of the food chain—bacteria and algae forming the diet of many microscopic organisms, the prey of larger animals, that are themselves the prey of larger carnivores. The food chain continues until one arrives at the top predator in the system, in this case our eurypterid and the larger carnivorous fish.

The ground was softer now, and the shore more muddy, so we picked our way with some caution. We had to cross small pools a few centimetres deep, and shallow inlets at the lake edge. Bob was leading the way and was about twenty metres ahead when he emitted a whoop of delight.

'Quick, come and look at this.'

Catching up rapidly I followed Bob's eagerly pointing finger.

'There, two of them, swimming away from us.'

I was nearly too late. All I could really see were two wakes in the surface and a muddy trail in the water.

'What are they?'

'Real live cephalaspids,' announced Bob with a grin. 'They were right at my feet, I nearly trod on one. I think they were rooting in the mud. When you can see the bottom it looks churned up and there are a few curved lines on the surface of the mud, maybe fin or tail marks.'

We walked cautiously through the shallows in the direction taken by our quarry. We were being overtaken by enthusiasm here, and the water was nearly up to our knees. It sounds really wimpish but we had to turn back; we were not equipped for wading. Reluctantly we turned parallel to the shore to gain shallow water and dry land, but we were stopped in our tracks by rapid movements in the water along the shore.

'Hey, look—a fish jumped.'

'Not a trout!'

'There's another, and another.'

'They are very small.'

'Yes, I bet they are *Mesacanthus* again.'

'They are coming this way. Stand still.'

As we stood in the water halfway up our calves, the fish disturbances in the water came nearer and nearer; occasionally small fish left the surface in twos and threes, but they were not feeding on anything on the surface; there were no hatching nymphs or mayflies in the Devonian.

'They are being chased by something. It is a shoal of small acanthodians being chased by something bigger.'

Now the shoal was only a few metres away and a larger swirl appeared in the water, causing small fish again to scatter. Then we saw the flashes in the water, the small ones of the shoal fish as well as the larger iridescent flash of a larger fish turning rapidly in the water in pursuit of its small and terrified quarry. The larger fish looked big through the water—more than thirty centimetres long; but this is part of the delusion of fishermen and the reality of refraction of light. They were more like twenty centimetres long.

It was not easy to see the details for identification, but occasionally one would break the surface in its chase, and we were convinced they had two dorsal fins. The best guess on identity was *Ischnacanthus*; it is the most common predator found in the quarry at Tillywhandland, and it had good strong jaws and teeth. Another bit of fossil evidence is that the fish bed contains a lot of fossil coprolites. Coprolites are fossil excreta and can provide useful information on the diet of the producer.

Many of the coprolites from Tillywhandland contain masses of scales and spines of *Mesacanthus*, and frequently all the spines in the coprolite point the same way. This is caused by the predator swallowing its prey whole, and head first. The spines retain their orientation on their passage through the gut of the predator—just as well, considering the sharpness of the points of the fin spines of *Mesacanthus*. Swallowing a *Mesacanthus* must have been rather like trying to swallow a modern stickleback, and many modern predators, like the pike, do not succeed in that exercise, spitting out the prickly prey. Whether by accident or design, goldfish in my pond sometimes get a stickleback stuck in their mouth and have to be rescued. I do not think the goldfish attack sticklebacks; they probably go for the same lot of food—goldfish opens mouth and sucks—in goes head of stickleback and it is stuck fast by the pectoral and dorsal fin spines. If the couple died and were preserved in the rocks, the resulting fossil would certainly be cited as evidence that goldfish preyed on sticklebacks.

The fish we were watching showed no fear of the two large erect animals standing in the water, but as soon as a foot was moved or even waved above the water, the fish shot off. Strange in a way; what could have produced this reaction since there was nothing on land or in the air that could have been a threat to them? Maybe just an inbuilt defensive reaction to escape from any movement they could not interpret. The converse can be seen in a pond with tame goldfish that approach a human being standing at the pondside, in the expectation of being fed. Fish are not stupid, and can learn a little by experience. However, I am not sure that my goldfish can distinguish me from a heron.

The *Ischnacanthus* had gone and we sploshed our way back to the lake shore. We were both in good humour, pleased with our observations that had generally confirmed the tentative interpretations we had made from the fossil evidence so patiently accumulated from the rocks by Bob. On the other hand there had not been any big surprises. The virtually non-preservable delicate plant, algal and bacterial material we had seen was perfectly familiar, and much would not have appeared out of place beside a modern, muddy lake shore. The difference here in the early Devonian was the absence of large plants, the silence in the air, not just through the lack of birdsong but also no rustling of leaves in the breeze.

We were walking back to the Bus, quietly contemplating all that we had seen, when Bob noted the first puff of ash. There had been plenty of small clouds hanging about and the top of the volcano had been at about cloud base for the duration of our excursion. Now there were fewer clouds, and a distinct greyish plume was rising gently from the volcano.

'Oh, look. Subduction-related andesitic volcanism.'

'Very good, but it's not doing much. It could have been smoking for hours, and we would not have noticed. We have been looking down all the time.'

'True, it was probably in the clouds.'

So we turned our backs on the volcano, and headed towards the Bus. We trudged back through the mud, past small pools, always on the lookout for life, pausing frequently to peer into the clear, shallow pools.

It was almost imperceptible, just a slight movement, and concentric ripples spread simultaneously on the pools around us. The volcano still looked the same, just a wee plume on top, but that had been a small tremor.

'Hey, my first earthquake, and in the Devonian.'

We had not gone far when it happened again, but stronger. Sand slopes in a pool collapsed in a flurry, and larger ripples spread over the surface; small waves lapped on the lake shore and spread thirty centimetres beyond the shoreline. The volcano still only smoked, but there seemed to be more than one source now. Small plumes issued from points on the flank of the volcano, but it was a long way off, and the air hazy, so it was difficult to be sure.

We had nearly two kilometres to go to the Bus. A twenty-minute saunter should do it.

Suddenly a booming roar made us spin round and face the volcano.

'She's blown!'

A column of billowing, black ash was rocketing into the sky. Explosions came thick and fast as the cloud built into the sky. To us it looked as though we were looking at a vertical column, but soon it was apparent that the wind was in our direction—at very least we would soon be showered in fine ash, and maybe worse.

It was a brisk scramble to the Bus. It was easy picking a path through mud, sand and shallow pools, but a totally different ball game when trying to run. The cloud of ash loomed ever closer. Luckily there was a gentle breeze; unluckily it was coming our way.

The explosions were increasing in intensity, and the whole volcano appeared wreathed in ash since we were looking into the cloud that was progressively blocking our view of the action. A vast column of ash towered into the sky; it was getting darker and flashes of lightning shot through the clouds. We were still 100 metres from the Bus when an anxious backward glance revealed a billowing cloud of ash pouring down the side of the volcano. We ran—there is no point in watching a pyroclastic flow. We could not see what was coming. We were not watching any more—just sprinting as best we could on the muddy sand.

By the time we had only fifty metres to go, there was debris in the air. It stuck in our gasping lungs, but at least it came from above, from the ash plume. We would already be dead if a pyroclastic flow had caught up with us. Stumbling into the Bus, we hit the starter. The motors fired and we were off, but this was no ordinary ride: we were pitched sideways and buffeted by impacts. There was only rushing greyness outside, no sense of place or direction—a real white-knuckle ride.

It seemed an eternity, but it was only twenty-one seconds before we shot into the light again. We had come uncomfortably close to the top of the pyroclastic flow, and had nearly been burnt up. Just as well the Bus was made with a heat-resistant skin. It was only on later examination of the outside of the Bus that we realised what a close call it had been. There were signs of scorching. We had touched the edge of the hot incandescent ash flow and lived; we had nearly become a clast in a welded tuff—not a nice way to go!

'Well, that's enough subduction-related magmatism for one day.'

'Yes, but we must search for a welded tuff in Strathmore and dig underneath, we might find that *Zenaspis*.'

Excursion 2

Hot Springs and Geysers

TIME: Early Devonian.

LOCATION: Rhynie, Aberdeenshire.

OBJECTIVES: A visit to the ancient hot springs and geysers of Rhynie, and examination of an early land-based ecosystem.

THE MODERN EVIDENCE: The superbly preserved biota of terrestrial and freshwater animals and plants that have been described from the Rhynie chert, and the associated deposits of a hot-spring system.

The Rhynie chert is a hard, silica-rich rock, similar in appearance to the flint from which Stone Age man made his tools. The chert at Rhynie was discovered in 1912 by William Mackie, a doctor in Elgin. He wrote a number of geological papers, and is recognised as a pioneer in the study of heavy minerals in sandstones. Mackie found the chert whilst making a geological map of the area to the west of Rhynie, the oft-repeated story being that he noticed this unusual rock in a wall he was sitting on whilst eating his lunch.

Mackie recognised fossil plant stems preserved in three dimensions within this rock, and made thin sections of the chert. When viewed under low magnification these revealed that the cellular structure of the plants was perfectly preserved (Fig 2.1). He had found the oldest well-preserved land plants on Earth, and from that time to the present day palaeontologists have studied the fossil flora and fauna of the chert in great detail. To date seven different spore-bearing plants, together with fungi, algae and the oldest-known lichen have been described from the chert.

There are also arthropod remains in the chert, which include crustaceans that lived in freshwater pools and land-dwelling, spider-like animals (trigonotarbids), the earliest harvestman spider, the earliest insects, mites, centipedes and other primitive arthropods. This is one of the oldest, well-studied fossil terrestrial ecosystems on Earth, and is highly unusual because the chert originated as siliceous sinter from hot-spring eruptions. Parts of a geyser vent have been found, and the textures in the chert match those seen today in the siliceous sinters deposited by

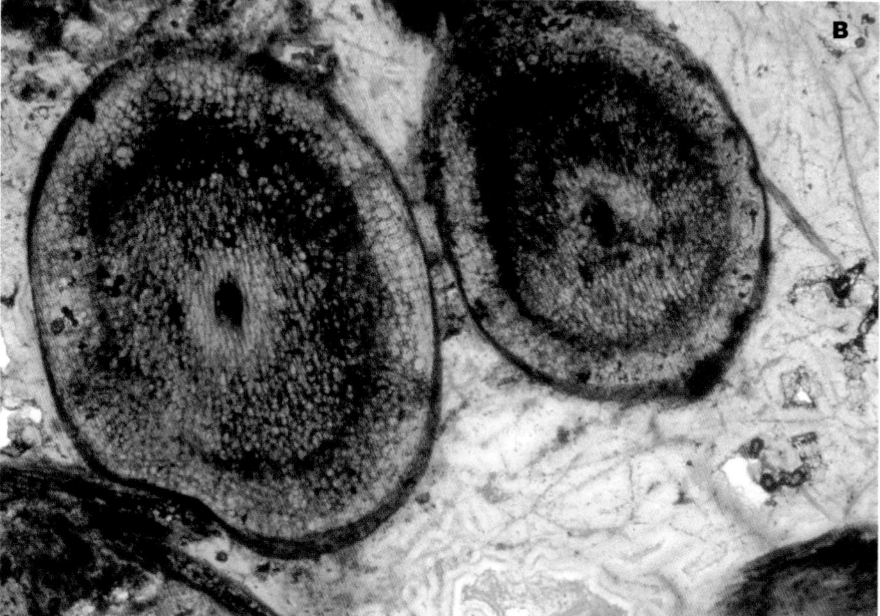

Fig 2.1 **A:** Cut-and-polished, vertical section through a bed of Rhynie chert showing the stems of *Rhynia* (darker lines) preserved in growth position. The weak horizontal layering is formed from bacterial mats that helped bind the plant stems together and prevented them collapsing. **B:** Thin section of a cross-section of *Rhynia* stems, 2–3 mm in diameter, showing the excellent preservation of cell structures within the plant.

geysers and hot springs in geothermal areas such as Rotorua in New Zealand and Yellowstone Park in the USA.

Lyall, my accomplice on this excursion, I have known from his days as a student in Aberdeen, and latterly as a member of a research group working at Aberdeen University on the geology and fossils of the Rhynie area. We have spent many days together at Rhynie, and in the more exciting surroundings of Yellowstone National Park with its hot springs and geysers. We had given ourselves the task of finding this small area of ancient hot springs in space and time, and examining the classic terrestrial biota of 410 million years ago. This would not be easy, and we would need to use the remote sensing facilities on our time-travel Bus. Giving due allowance to the shifts of geography that have taken place since the Devonian we fixed co-ordinates and height to arrive at 10,000 metres over the Rhynie at the appropriate time.

The view on the screens was not entirely as we expected; but then, it hardly ever is. The land rose to the south into a mountainous region with about a kilometre of topographic relief, but large areas were obscured by cloud. The land sloped northwards towards the region that is now the Moray Firth, and there was surprisingly little rock outcrop to be seen. A few hills poked through the sediment cover and north–south alignments of land features such as hills and valleys were apparent. Sunlight reflected off water in rivers draining to the north and from the surfaces of lakes that were dotted along the valleys. Far to the north there was no sign of bedrock, just a vast area of river channels, lakes and sand flats.

'Basically this is OK, but there is a lot more sediment than I thought. The Grampian Block is well covered and it doesn't look as though there is any major topographic relief.'

'Unless it is all buried, and we are in the wrong timescale by a few million years,' ventured Lyall.

'Yes, that's possible. The Rhynie cherts are only a couple of hundred metres above the basement of Dalradian metamorphic rocks and Ordovician igneous intrusions, so we don't want to see too thick a cover.'

'How do we pick a landing in that lot? Rhynie could be anywhere, and I can't see any hot springs from this distance.'

'We will have to use remote sensing and see if anything turns up. There was a volcano at Rhynie just prior to the deposition of the Rhynie cherts, and the cherts were deposited from hot springs, probably as the volcanism died out. The first stage then might be to look at a surface heat map and see if we can pick up any hot spots. We could also try for emissions of gas (CO_2, H_2S) from hot springs or volcanoes, but this would require actual eruptions to be taking place, so we would have to be right on the button in time and space.'

'How big is the footprint we can cover with the heat map?'

'Only about a ten-kilometre diameter from this altitude, so let's take the Bus up to six kilometres for a better view. Then we can cover a thirty-kilometre diameter and should have a better chance of finding the spot.'

As we ascended the view widened out, and a pattern of cloud over the mountains became apparent. A band of cloud stretched to the southwest, and beyond the cloud there was clear sky. There was a sharp margin to the mountains and this looked like it might mark the line of the Highland Boundary Fault.

The Highland Boundary Fault divides the Midland Valley of Scotland from the Grampian Mountains. In the early Devonian it was uplifted on its north-west side, probably due to granite intrusion at depth in the Grampian area. The fault can be seen today at Stonehaven, where it separates the metamorphic rocks of the Grampians from the Old Red Sandstone of the Midland Valley, and forms a prominent northeast–southwest line on the geological map (Fig I.2), and also controls the present-day topography of Scotland. Importantly, we had found the Grampian Block in which our objective lies. Rhynie must be on a line thirty kilometres north of the fault, and we could now restrict our search area. Since we were looking for hot springs the easiest option was to use our remote heat sensor and map ground temperature in the area.

The heat scan took several minutes to build on the screen. Colours produced on the scan reflect temperature and range from blue for cool through green and yellow to red for hot. Our scan was mainly blue reflecting cool ground temperatures, but there were also flashes of yellow and an interesting small red spot to examine. The spot lay in a valley close to a river, and we could detect that the water temperature in the river increased by several degrees where it passed the hot area. The visual scan of the area showed a valley with a sharp western margin. Bedrock was exposed to the west and sediment clothed the valley floor, the straight margin to the valley implied that it was fault controlled. So far so good, this could be the Rhynie basin that contained the hot springs we were seeking.

As a confirmation we checked variations in surface chemistry in the area, discovering that a broad area of dark rocks that crossed the valley was low in silica; these appeared to be basic igneous rocks. This was very exciting, and we soon deduced that the flood plain was hotter over an area some two kilometres long by 500 metres wide, and the river 4°C hotter where it left the area than at the entry point.

'That's it, we have found Rhynie. A warm region where a basic intrusion crosses the Devonian valley, it's almost like looking down on the geological map.'

We spent a long time making detailed remote scans of the area at slightly different times in order to pin down our landing location. Slightly earlier in time we had found a small active volcano in the area, and the hot surface temperatures in the sediments seemed to follow the end of volcanic activity. Eventually we were satisfied with both time and location. By then it was too late in the day to land, so we would have to wait till dawn.

The hour before sunset was a time to relax; time to consider the prospects for the morrow. Gazing from the ports of the Bus was a favourite occupation, the views spectacular and ever changing. On this Devonian evening as the light began to fade the hills were bathed in pink as clouds darkened to purple and mauve and the sky became yellowish and then bright red—a brilliant sunset. The only time I have seen such a brilliant red sky was in Western Australia in 1982, at a time when there was a lot of dust in the upper atmosphere from a recent eruption of Mount Pinatubo in Indonesia. It seems logical that we had the same effect here; there was a lot of volcanic activity at this time in the Devonian and chances are that a volcano was active somewhere in the Scottish area. Another factor may have been the amount of dry sediment on Earth, the atmosphere may just have been generally dusty with material carried into the upper atmosphere by storms, in much the same way as dust from the Sahara sometimes reaches Britain and falls in rain droplets, usually when the car has just been washed. The sun set rapidly as a deep red disc; we were only 25 degrees from the equator and there was little twilight. Darkness soon enveloped the scene.

It was an early start, three hours before local dawn, with all the technicalities of preparing the Bus for the day's activities. This was a look-see excursion, no materials to be collected and strict biological controls to ensure that no bacteria are brought back to contaminate present time. Nobody could predict the behaviour of a 410 million-year-old bacterium, so it was deemed best to give them no chance to demonstrate any adaptability to the present. Some bacterial epidemics had occurred in an American time-travel facility, and no chances were taken. Even politicians had woken up to the danger, following fallout from mistakes made in experimentation with genetically modified crops in the early years of the century. Furthermore we were visiting a hot-spring area where heat-loving, 'thermophyllic' bacteria probably existed, and many of these forms can thrive in extreme chemical as well as thermal environments. Whilst life itself probably evolved in such a form, it is unwise to re-introduce such a organism to a new set of environmental parameters, and allow it to interact with its progeny so much further down the line.

We landed the Bus on the eastern side of the river beside an outcrop of dark weathered rock with pinkish patches and pale veins, probably part of the Ordovician basic igneous rocks that intrude the Dalradian in this area (Fig 2.2). The surface was crumbly and small gullies carried the rocky debris of weathering

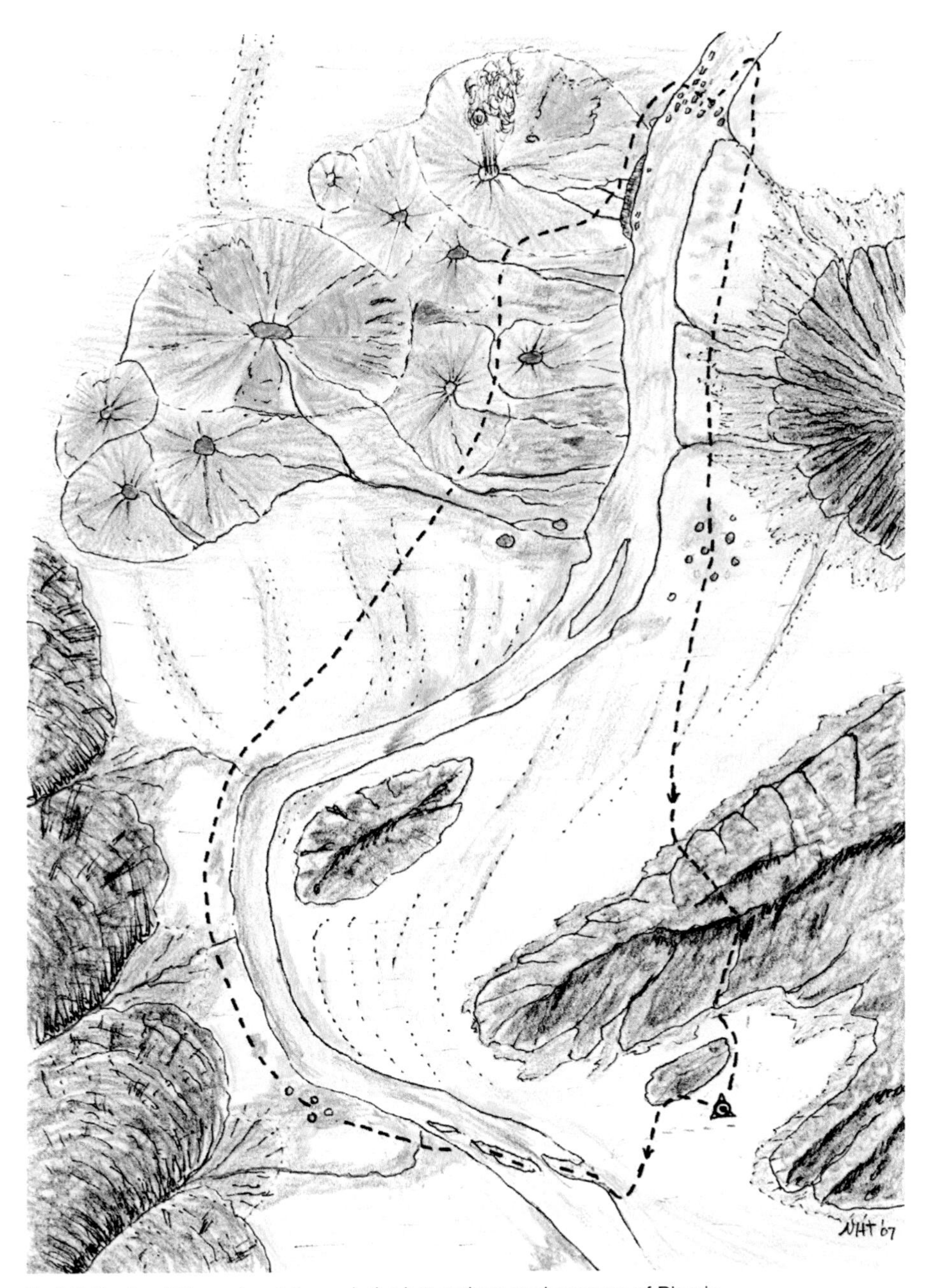

Fig 2.2 Route of Excursion 2 through the hot springs and geysers of Rhynie.

to gravelly spreads that merged with the river floodplain. A river with channels dividing around sand bars occupied the western side of the plain. The water was clear and reflected the blue of the sky, but more exciting were the patches of green that dotted the river plain—plants, and lots of them. Beyond the river lay a line of hills with a scarp slope facing us, the scarp cut by valleys that ended in small alluvial fans of sand and gravel. This could reflect the line of the fault that bounded the basin. We had the privilege to wander the banks of this ancient river and examine the vegetation and search for beasties.

We emerged from the Bus into a pleasant early Devonian morning. The temperature was 15°C and clearly heading higher as the sun rose in the sky. A gentle breeze blew from the north, and that was about it. Silence; the silence was so uncanny to anyone used to country life. There was no birdsong, no trees to rustle in the breeze, no animal noises, and best of all no traffic. The river provided most of the noise as it swept past sand bars splashing and rippling against the banks.

We were drawn first to the rocky outcrop area beside the landing site. The dark crystalline basic rock, probably a gabbro, was rough and weathered. A few paler and more resistant veins stood proud of the surface; they contained quartz and were more resistant to weathering.

'What's this?', ventured Lyall, pointing at greenish patches on the rock, 'Looks like lichens to me. They are all around us.'

Sure enough there were clearly several types of lichen colonising the bare rock surfaces. The earliest known lichen comes from the Rhynie chert, and that was clearly a lucky accident of preservation. It is so unlikely that bare terrestrial rock surfaces and their encrusting lichens become preserved that lichens with such habits have virtually no geological record.

'Let's inspect your river. It looks as though we will have to cross it to see the best vegetated areas. Let's hope it's not too deep.'

Between our landing site and the river the surface was mostly sandy, with patches of gravel. Clearly the river flooded from time to time and washed over this area, leaving relics of shallow flood channels. Soon we were standing on the bank of a shallow river about fifty metres wide. Water rippled over a pebble bed between the bare sand and gravel bars that emerged from the river channel. However, there were deeper pools where the current had eroded the banks, and it was into one of these pools that we were looking. The water was about a metre deep and clear. There was little sign of movement apart from green filamentous algae forming streamers on the pebbles.

'Can't see any fish.'

'Just as well,' said Lyall. 'That would be an embarrassment since you have worked on the Rhynie area for years and never found a fossil fish.'

'At least fishing experience is some use. You learn the shapes of sand bars better from wading than from reading John Allen's sedimentology texts, and it doesn't involve equations. We will need to take some care as the water is faster in the shallows and the gravel can wash out from your feet. There are no early Devonian trees to provide a wading staff.'

From the tail of the pool the crossing to the sand bar was uneventful, splosh-ing through water up to our knees and sending flurries of sand downstream as we disturbed the sediment, but no obvious sign of life. We walked out on to the gravelly head of the bar and noted the deeper faster water on the far side, but it looked like we could cross about fifty metres downstream. We scrunched over the pebbles and on to the sandy part of the bar. The top bar was smooth, but ripples took over on the lower regions and there were ridges parallel to the water's edge, recording a recent fall in water level. The river level obviously fell quite fast following a spate. We had seen the clouds to the south yesterday from the Bus, but today was all clear blue sky, and there seemed little danger that we would be trapped on the wrong side of the river by a flash flood.

We wandered separately over the sandy bar, fascinated to see the sorts of struc-tures that we knew from cross-sections of rock cores we had taken during our drilling investigations at Rhynie. A whoop of delight from Lyall signalled our first animal evidence. I walked over to find Lyall proudly surveying a trackway in the sand.

The trackway consisted of two parallel rows of closely spaced leg imprints, looking like the mark of worn bike tyre. The beast had wandered along the edge of the water, apparently being happy to leave water and return to water. The bizarre thing was that it could instantly be given a trace fossil name—it was *Diplichnites*, and a very good example, as it should be since it was made in the last day or so and not 410 million years previously (Fig 2.3).

The trackway was generally gently curving, lacking sharp turns, but where it did curve the individual leg imprints could be seen in rows of about 15–20 distinct marks. This showed that our animal had many pairs of very similar legs and was some form of myriapod, allied to our modern millipedes and centipedes. However, this animal had a leg span of some fifteen centimetres. The tracks ended at the water's edge, but considering that water level in the river was falling, the beast could not be far away—was it in water or had it swum or crawled to the other side?

Fig 2.3 *Diplichnites* trackways made by large arthropods preserved in sandstone deposited by a river. (Lower Old Red Sandstone, Quarry Hill, near Rhynie.)

Whilst tracks of this kind are well known from Silurian and Devonian strata we do not know whether the animal responsible was largely terrestrial or aquatic. We had evidence here that it could walk out of water and support its own weight: no belly drag marks in the sand. Equally, it was apparently at home in water. Did it lay its eggs in water or on land? Did it breath air or have gills or both? Those and other questions rushed to be answered, but first of all we had to find the animal.

Judging from the track it had to be at least thirty centimetres long, and probably the largest beast about, so where could it hide? It was not still wandering around our sand bar, so we waded carefully through the shallow water at the tail of the pool and headed for the far side of the river. We trod gingerly, testing the sand—it would not be a good idea to tread on one and it could be hiding in the sand. In the event, our feet disturbed nothing and we reached the main bank, a cut bank about fifty centimetres high and made of laminated and rippled sand being eroded by the lateral migration of the channel. The bank was hard sand, and firm to walk on; there was clearly some form of binding agent or cement. The river flooded the surface periodically and had left streaks of lighter-coloured sand, together with stranded pebbles and plant debris deposited at the high water mark of the last flood. Some pieces of vegetation were remarkably large—up to thirty centimetres in diameter and nearly two metres long.

'Do you really think these logs could have come from trees? I didn't think there were plants this size in the early Devonian.'

'I don't think there were; we need a palaeobotanist here. It might be a log of one of the problematic things called nematophytes. *Prototaxites* and *Nematoplexus* are names applied to pieces of these 'plants' that are found in the Rhynie chert. The names are based on microscopic structure; we do not know what the whole object looked like. The vegetation looks rather fibrous.'

A desultory kick dislodged a fragment; it was certainly soft and broke down easily. Under a hand lens a mass of knotted tubes could be seen with finer tissue between. There was a distinct nasty smell to it, reminiscent of mushrooms. Scientific opinion was divided; some took *Nematoplexus* to be some kind of fungus or alga that lay flat on the land surface, or maybe grew in the water. It did not seem to be one of the more advanced plants, types of which were still small in the early Devonian.

'It has a round stem, so I don't see why it should grow horizontally—it looks as though it ought to stand up; it certainly looks strong enough.'

Nearby, smaller examples were dried and brittle, easily breaking into small fragments between the fingers. Clearly it dried out and disintegrated or was decomposed rapidly.

'Well, if it grows upright we should be able to see some sticking up like pillars, and I can't see any at present, so maybe it has been transported some distance.'

'Possibly, but remember there is a fossil sand-filled cast of one of these from the Rhynie area which was apparently preserved in an upright position—it's in the geology department at Aberdeen University.'

This was an unsatisfactory end to the conversation: no conclusion. Lyall kicked the log again, and it rolled a few centimetres from its sandy resting place. Dry sand trickled into the resulting hollow, but there was another movement; the sand was being heaved up from below.

'Something is under there.'

'Yes,' muttered Lyall.

'Is it going to come out in the name of research or do you have to poke it?'

'Why me?'

'Because you are a friend of all arthropods, and it might be a scorpion.'

'There are no sticks,' Lyall pointed out.

'Try a bit of *Nematoplexus*.'

Lyall gave in, and poked the sand gently. The sand moved again and a small head with two antennae nosed out of the sand, paused and then continued to emerge. It became wider and longer with flat segments from which legs protruded. It was rather like a giant woodlouse, but with more obvious subdivision into segments. Our beast was not a fast mover, its legs moving slowly and deliberately with a rhythm that passed in waves down its body.

This was our trackway maker, hiding up under the *Nematoplexus* log. It trundled off over the hard surface, leaving no recognisable track except where it crossed dry sand. It moved purposefully, probably looking for a new hiding place. We walked with it, at a respectful distance; after all we were the biggest animals it had ever seen.

'Count the segments,' ordered Lyall.

We both made it twenty-three plus a head.

'Turn it over! We need to know the leg arrangement.'

We soon discovered it was a stable animal and did not want to roll over and have its underside examined, but a deft flick with a bit of *Nematoplexus* had it helpless on its back, and trying to roll up in an attempt to right itself.

'One pair of legs on each segment, legs with about ten, yes, ten segments and a double claw at the end. No legs on the last segment, mouth parts look like blunt munchers, probably a general scavenger or detritus feeder.'

The legs flailed their metachronal rhythm to no effect in the air. It was not used to being upset and had no means to right itself. Lyall flipped it back over and it set off at double pace. We watched it head towards the water and felt we were going to see a swimming demonstration when our friend just disappeared into the ground.

'Hey! It's escaped. Damnation.'

On arrival at the dematerialisation site the solution was obvious. Not a case of 'beam me up Scottie' but rather the arthropod had gone to ground. A neat burrow entrance some ten centimetres in diameter pierced the sand, and our friend had disappeared down it.

'Hmm, I suppose the burrow is, or might one day become, *Beaconites*. It can't have had much room to spare.' As if to emphasise the point a scuffling noise came from the burrow and a pile of sand emerged to be flicked away from the burrow entrance by the flat, legless tail segment.

'It's doing a bit of housework.'

'It's been doing something else as well, look here.'

Ovoid pellets about five millimetres long littered the area around the burrow. Coprolites—or as we see them just faecal pellets. Closer examination showed them to be made of macerated vegetable matter; the arthropod appeared to be a vegetarian.

This was excellent. We had proved the connection between *Diplichnites* trackways and *Beaconites* burrows and met the maker of both. All this reminded me of the Townsend Tuff in the Old Red Sandstone of South Wales. John Allan and Brian Williams described a surface overlying a sandstone bed with *Beaconites* burrows, and the top of the surface was strewn with coprolites. In that case the surface was preserved under a blanket of volcanic ash, and presumably the animals were not too happy about the volcanic eruption. The facetious suggestion was made (but not in the authors' paper) that the animals were severely frightened when they saw the volcanic ash cloud coming. Also in south Wales, at Tredomen Quarry, Phil Bennett found a fossil arthropod more than thirty centimetres long and with at least ten segments. This may well be the animal that made *Beaconites* burrows and *Diplichnites* trackways; both these trace fossils are found in the same quarry.

We needed to press on and find the green patches we could see from the hill-side. Following the river downstream we soon found patches of plants growing in damp sandy hollows along the river banks—areas where dampness remained after river floods and from runoff from gullies cutting the scarp face to the west. The surface here seemed to be close to the local water table. The plants were low, none exceeding thirty centimetres in height, and they grew as expanding patches which eventually joined together to form a green sward of upright stems. The clumps expanded by runners on the surface, or just beneath the surface. From the runners arose vertical leafless stems of pale green. Some stems ended with small torpedo-shaped capsules. Each individual plant patch was about thirty centimetres across, and it appeared as though they were intent on colonising as much of the surface as possible before others had the chance. The capsules on the stems were varying shades of green and brown, the brown ones near the centre of the clump, and younger, less mature green ones nearer the margins. The green ones were soft and moist, but the brown ones were dry and brittle and cracked easily in the fingers to release a small cloud of white dust—the spores of the plant. These were typical sporophyte plants, growing rapidly and producing countless millions of spores. A few of these spores would fall or be blown into a suitably damp area where they would develop into small gametophyte plants—the sexual stage of the life history.

The clump we were looking at appeared to be *Rhynia*, one of the commonest plants in the Rhynie chert (Fig 2.4). Nearby were different forms; one with bulbous bases to the stem and a many branched stem was clearly *Horneophyton*, and a rather straggly looking plant with stems up to five millimetres thick and arched basal runners conformed with ideas on *Aglaophyton*. The plants competed for space at the edge of the clumps, and the stems of different plants mixed to form a miniature green forest of stems and runners. Many much more advanced

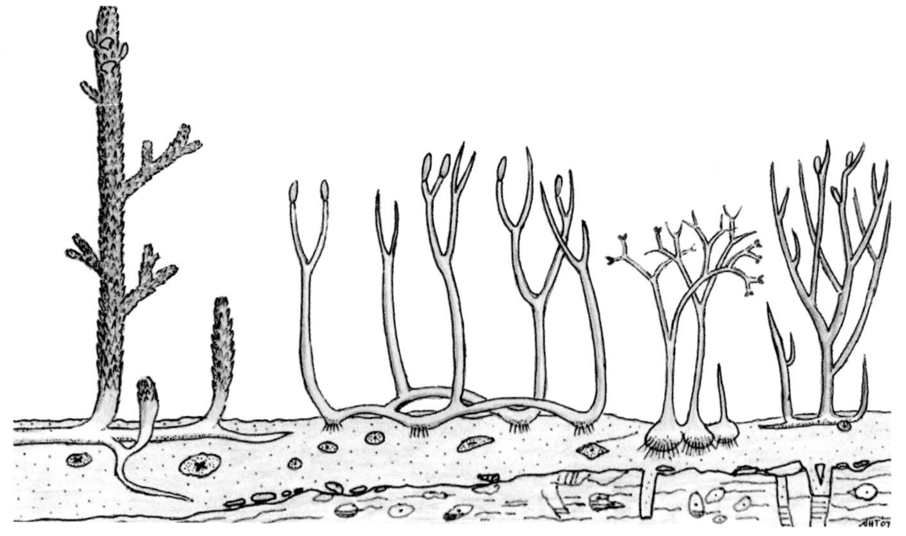

Fig 2.4 Sketch of plants found in the Rhynie chert; from left to right: *Asteroxylon*, *Aglaophyton*, *Horneophyton* and *Rhynia*. The largest plant was about 30 cm high.

plants do the same thing today—just think of the vegetative spread of chickweed or clover from the small parent seedling. There were no dead plants here; they must have all grown up together as the initial colonisers of the sand deposited by a previous flood, or maybe their spores were blown into the hollow. There was no obvious subdivision of the area into different habitats, or any apparent preference on the part of the plants for growing areas or neighbours.

'That's a bit of a shame. Some of us have speculated long and hard on the influence of organic content of the soil, the type of substrate and the availability of water to determine orders of succession for the plants, but here we can pick a bunch of Devonian vegetation with four or five different plants all growing on sand and in a small area. At least it shows why lots of chert beds only have one or two plants—if you imagine a borehole anywhere here it would only sample part of one clump, or the small area where two or three clumps interfere. Our boreholes through the Rhynie chert can't give a representative picture, but they are much better than relying on blocks dug from the field.'

Our bunch of Devonian vegetation now had five representatives, including *Asteroxylon*—a plant with small scaly bracts covering the stem, and kidney-shaped sporangia nestling between the bracts near the tips of the upright shoots. The bunch was not very colourful, just greens with a bit of brown. The *Aglaophyton* was already going limp; it has no spiral thickenings in the central strand of the stem. Clearly *Aglaophyton* is more reliant on water to hold it up; it goes limp very quickly when picked. Dianne Edwards, who has described detailed structure in the Rhynie chert plants and tried to educate us on the plants at Yellowstone, would love this plant collection. Sadly we cannot take our posy home for her.

Leaving our bunch of vegetation behind, like a small memorial for the Devonian, we headed on down the river, debating the names of the plants as we subconsciously tried to walk between them rather than squash them—no need to alter the future of plant evolution by a careless step. We saw several plants that would be new to science. This was not really a scientific surprise because we know from Charles Wellman's work on spores from the Rhynie sediments that many more types of plants existed in the area, and the fossil record is always incomplete. Our wanderings took us away from the river channel and the background noise of the stream died away towards the characteristic Devonian silence.

We crossed a dry stream bed where sand ripples had been left from the last time water flowed down the channel. On the sand were clear trackways up to a centimetre wide and made by small arthropods. The trackways had a patter of tiny footprints either side of a central intermittent groove. As a trace fossil this would be called *Siskemia*, but we were no nearer knowing what the animal looked like than we were when examining the fossil version. No recorded arthropod from the chert had the required features to produce this trackway. There were also sinuous trails looking like those made by worms crawling over mud. We could not spend time hunting for the beasts now; maybe later if we made good time.

Pebbles in the stream had been washed from the hillside and revealed slaty fragments and metamorphosed grits. These were instantly recognisable as fragments of the Dalradian metamorphic rocks that form much of the hills to the west of the Ugie valley to the present day. These rocks clearly differed from the weathered gabbro we had seen back near the bus; we have now crossed the line of the fault noted from the remote sensing before we landed.

We ate our lunch sitting in a shallow hollow, surrounded by plants growing from a sandy surface. Plant litter of dried stem fragments was trapped amongst the plant stems and creeping rhizomes. Elsewhere the bare sandy surface was swept clean, probably by breezes blowing the light debris away to accumulate in sheltered areas, just as the wind blows today's autumn leaves.

Our Devonian day was calm, and we could peer closely at the plants and the underlying litter. We knew there had to be animal life here, but the problem was spotting it. Virtually all the arthropods known from the Rhynie chert are less than a centimetre long. This is possibly because we look at the chert using thin sections and use a microscope, so we could easily miss larger animal fragments.

Nothing met our gaze, so patience rapidly diminished and we started prodding the litter in the hope of waking something up, and frightening it into action. The first discovery was not animal, but fungus. The damp litter was irregularly stuck together with a delicate network of pale hair-like threads that covered plant debris and stretched mesh-like across cavities. The mesh could be seen inside damp hollow straws of partly decayed plant stems. We clearly had a fungal presence that was breaking down the plant debris and releasing nutrients to be used by the growing plants—maybe the oldest compost heap.

'Ah, look. Under that stem I can see a pair of antennae. Something is waving at us out of the end of that dried-up straw.'

'Yes, I can see it—will it come out?'

'It's got a choice—come quietly or be pushed.'

We decided on the 'push' method, but with the cunning expedient of the sample bag trick. Pick up the straw quickly and pop it straight into the bag. This was somewhat unfair as the arthropod never saw us coming. It soon became rather upset. Two dark brown antennae, just a few millimetres long, waved from the end of the straw. A rounded shiny brown head appeared and a pair of segmented legs. Our animal then went into overdrive, and shot out of its hiding place in the dried stem.

'Fantastic. It's a centipede.'

Now this might not seem the greatest excitement imaginable, but it was a most welcome sight for us. Back in 1999 Lyall and I had found some bits of an arthropod in the chert which we had described as parts of a centipede. It is always risky to identify an animal from bits, especially when comparing features of modern centipedes with their ancestors 410 million years previously. We had found a segmented structure with twenty segments, hairs, and a spade-like end segment. We had considered various options. Was it a wee hairy worm? Was it a tiny hairy millipede? Close investigation eliminated those options, the second being eliminated when, after carefully isolating the specimen in a slice of chert and grinding the rock down so we could see the other side of the object, we determined that it had no legs. Whilst cutting the specimen from its block of chert we also found a leg segment; this closely resembled that of a modern centipede, and then the penny dropped. A weekend at home in the garden produced a fine selection of centipedes of various types, and the structure of our fossil closely resembled a centipede antenna. We had published our results, but there was always the nagging question: what did the whole animal look like, would it be a centipede? Now we had our answer.

The prisoner in the plastic sample bag was distinctly annoyed, running rapidly about, but having trouble on the slippery plastic. A quick inspection revealed a strong pair of jaws under the head; it was clearly a carnivore, as are others of the Rhynie arthropods. Lyall tipped it back into the plants, and our centipede rapidly disappeared.

The *Aglaophyton* plants at our picnic site were mature and the sporangia, about the size and shape of a date stone, were dried and brown with many split open, having released the spores inside. A tap on a sporangium produced a shower of spores, which fell to the ground and mingled with the accumulating litter. Older sporangia had lost their spores but remained as dried hollow husks at the ends of the stems. The stems themselves were also drying out and shrinking, the outer surface becoming ridged and the colour yellow shading to brown. In fact, we could find all the stages of plant death and decay that are so perfectly preserved in the Rhynie chert.

We set up a series of plant fragments to photograph showing stages in sporangium development and decay, and while doing this we found our next animal. Out of the dried husk of a sporangium there crawled a small spider-like animal, but apparently with ten legs rather than eight, although two are actually feelers, and an abdomen clearly showing segmentation (Fig 2.5). It also seemed angry, and rather fierce for its size, by drawing itself up to its full ten-millimetre height and displaying a pair of oversize fangs.

'Looks like it was lurking in the sporangium, maybe as an ambush site, or merely as a safe resting place,' commented Lyall.

'What do you think it eats?'

'Well, there must be smaller arthropods—the mite and the springtail which we haven't seen, and are only a millimetre or so in size. They could also have cannibalistic tendencies of course.'

'Why don't we find another and introduce them to each other and see what happens—sex or mealtime are probable outcomes.'

However, we did not have unlimited time and another large *Palaeocharinus*—for that was its name—could not be found, although we did locate a group of four juveniles hiding together in another dried sporangium. The youngsters seemed to stick together for a while after hatching before becoming independent. A pity we could not introduce a couple. The whole scene would be reminiscent of stag-beetle fights staged in primary school playgrounds back in the 1950s and long before. All that was required was two male beetles and one female to bring out the competitive instincts in the males. Presumably this practice is now about

Fig 2.5 Model of *Palaeocharinus tuberculatus*. This is a trigonotarbid—an air-breathing, carnivorous arthropod, about 10 mm long, that hunted other arthropods on land. (Reconstruction by Steve Fayers, model by Stephen Caine.)

as politically correct and environmentally aware as cock fighting or bear baiting. Stag beetles are now protected animals, and much scarcer, pesticides rather than schoolboys being the main culprits in their demise. The schoolboys were merely aiding the process of natural selection.

Time was up for beastie hunting. We had to press on and find the hot area defined from the Bus. Were there active hot springs in the area? We had moved less than a hundred metres before we were distracted as the ground became wet and marshy, and the surface crunchy like breakfast cereal but with the pore space filled with water. There were a few small, roughly circular pools only a metre or so across, and we could not resist stopping and kneeling down to peer into the water. The pool was surrounded by stands of young *Rhynia*, bright green in colour, which squashed easily when knelt on and left nasty green stains on our knees.

Gazing into the small pool through stands of *Rhynia* a miniature ecosystem met our eyes. The pool was only about twenty centimetres deep and the water crystal clear. Plants toppled into the pool around its margins, and stems floated upside down in the water in various states of decay. A mat of dark grey-green

Fig 2.6 Model of *Heterocrania*. Remains of this arthropod, which was *c*.15 mm long, are usually associated with deposits from freshwater pools. (Model by Stephen Caine.)

slime coated the stems and formed regular pustules on the surface, grown over small spines projecting from the plant stems. There was a general slimy appearance to everything, but close observation revealed movement in the water. Tiny shrimp-like animals, only three millimetres long, were swimming through the water, the light picking up the rhythmic beating of their legs.

'Why are they swimming upside down?'

'Why not? They swim to feed and process microparticles picked up by hairs on their legs and pass them forward to the mouth—upside down is easiest because they can keep near the surface and avoid picking up the sediment below them. Basically they are filter feeding.' These animals looked like *Lepidocaris*, the most common of the freshwater arthropods found in the chert.

Further examination of the pool revealed a few different animals crawling over the slime at the bottom of the pool. They were no more than fifteen millimetres long with a simple segmented body, a rounded head and a sharp tail spine. Surprisingly they were bright blue. At first we were distracted by the colour, but soon realised we were looking at *Heterocrania*, a representative of an obscure and extinct group of arthropods called euthycarcinoids (Fig 2.6).

As we watched, a patch of the slime detached itself from the bottom of the pool, and buoyed up by bubbles it rose to the surface and drifted to the margin to join an area of scum on the surface. This was a bacterial mat, photosynthesising and producing oxygen. Gradually these scum surfaces can extend to cover small ponds, and can become thick and rubbery. Under warm conditions silica can be deposited in this floating mat by evaporation of water saturated in silica. This is one way that crusts are formed over ponds in hot spring areas, presenting a serious danger to the unwary, or the unlucky.

The surface on which we were kneeling, covered with a stand of *Rhynia*, was made of fragments of sinter deposited from hot springs together with debris from previous generations of *Rhynia*. The surface was loose with platy fragments of sinter a few centimetres across. Beneath the surface debris the sinter was hard and cemented. The walls of the pool we were gazing into were steep and encrusted in bacterial mat, but underneath the mat there was a hard, smooth surface. Our pool had once been a hot spring vent, erupting hot, silica-rich waters from which the sinter had been deposited. Some quirk of the local plumbing of the hot-spring system had resulted in the vent becoming blocked, and cut off from the hot water supply. The area had cooled and the vent had become a small pool, cool enough for bacteria, plants and animals to survive, and indeed to thrive.

How long had this taken? Was it a passage of years, months or days to effect the change from boiling hot spring to cool pool? Evidence from modern hot-springs shows that the transition from cool to hot can be virtually instantaneous; all that is required is a breakthrough of hot water. There are several possible causes such as small earthquakes, or build up of water pressure when a nearby vent has become blocked.

The transition from hot to cold and the colonisation by plants and animals are likely to be gentler processes. Various algae and bacteria can tolerate different temperatures, so the development of an ecosystem would depend on the rate at which the pool cooled. It would probably take a few seasons to build up an aquatic fauna. Aquatic animals would need to find the pool to lay eggs. Plants would have to grow to maturity from spores. The fact that there was abundant old, dried, debris of *Rhynia* from previous seasons indicated that our pool had cooled down at least a year ago, possibly several years.

This pool certainly looked an ideal candidate to compare with the Windyfield chert, in which we had found plants, bacterial films and animals both of aquatic and terrestrial origin. Our luck at Windyfield was that a plant and several animals previously unknown to science had either colonised or stumbled into our pool, giving us a window on early Devonian life. Contemplating this little pool was justification for our work on the Windyfield chert, and the description of its flora and fauna. The painstaking examination of slice after slice of chert had taken years, always expecting and hoping for a 'new' find, but maybe being disappointed for months on end, until a lucky cut revealed a scrap of material that could not be assigned to a known representative of the fauna or flora. The initial find was followed by exhaustive searching of literature, and checking of previous material, before coming to the conclusion that we had found something new. After this we needed to describe carefully, and scientifically justify our interpretation of the find. Yet here, it was all presented to us in a single view; we could watch the plants and animals, confirm some thoughts and cheerfully deny others.

Suddenly we were jolted from this introverted reverie.

'What was that?' A chorus broke the silence. 'I heard a roaring noise.'

'So did I, but I don't see anything.'

We walked on in silence, past patches of vegetation and over the beds of streams fed from gullies in the hills. We were both more attentive to the wider environment; plant and beastie spotting was abandoned. The river veered to the east around a bluff of basement rocks, and as we turned the corner a new vista gradually unfolded. There were low white mounds poking above the general level of the flood plain: some were streaked with orange, brown and greens of a hue we had not seen before.

We actually saw it, before we heard it; a column of water and steam shot twenty metres into the air from a mound about 500 metres away, continuing for over a minute with the roaring noise we had heard before. The cloud of water droplets and vapour drifted away on the breeze, and the sun's rays refracted to form a short rainbow in the residue of the geyser eruption.

'Eureka! We have a geyser at Rhynie and it's putting on a show for us,' Lyall observed with glee.

The eruption had come from one of the mounds, and we were naturally drawn towards the source of our Devonian geyser. We left the riverside and walked over the sandy river plain; it was dry here and few plants grew, but there were patches of dried plant debris left in hollows from the last flood. Before we reached the site of the eruption our progress was suddenly halted. We had come to the edge of a small stream, but this was no ordinary stream. A faint cloud of steam rose from the surface, and the bed of the stream was white with brown and grey streamers of fine filamentous strands waving in the current. At the margins a bright green line followed the margin of the stream. The green was an encrustation or coating on the grains at the stream margin, and the colour changed in a distinct zoned manner away from the stream. The sediment was different as well, being crumbly, crunchy, light white to grey granules, highly porous and soaking with water—warm water.

The ground was warm here. This was more than just a warm stream from a geyser; we were in an area of hot ground such as we had seen on the heat sensor from the Bus. Care was now the order of the day; hot-spring areas are dangerous. Boiling water can be very near the surface, and the rock and sediment can be cavernous, with pools roofed over with deposits from the spring to form a thin crust through which the unwary can fall.

'Hmm, yes, I have a friend who knows all about that, she put her foot in hot water in New Zealand once—not a good thing to do.'

'And Smokey did the same in Yellowstone at Norris Basin. That was into hot mud which stuck to his leg like an overheated poultice—nasty.'

Smokey was a Yellowstone Ranger of at least 95 kg and provided a good ground-bearing test. We always followed him when we were off-boardwalk and collecting research samples in Yellowstone.

This was outflow from a hot spring; it was not carrying any sand, only bits of sinter deposited by the hot spring. The sinter fragments were small and crunchy. Closer inspection showed some to be loosely cemented together with white silica. Amongst the debris were short, straw-like objects up to five millimetres in diameter; this was our first sign of preservation of the Rhynie flora. Stems had been encrusted with silica, and when the plant decayed the delicate straw remained. Most of the straws lacked any preserved plant material, but by breaking the straws an imprint of the outside of the plant stem could be seen preserved on the inside of the straw. After searching we found a few straws in which silica preserved the few outer layers of cells of the plant walls, but nothing like the perfect, complete silicification of plants seen in some parts of the Rhynie chert.

We had to cross this warm outflow stream because it clearly flowed into a marshy area near the river that included some hot vents and looked distinctly unsafe. Gentle testing soon revealed that the bed of the stream was rock hard—silica was being deposited from the hot waters on the bed of the stream. The stream was flowing over a broad outwash surface, and plants were less frequent, but here and there patches of *Horneophyton* grew on the bare sinter surface. There were also patches where plants had previously thrived, but now all that remained was the radiating pattern of their rhizomes on the surface of the sinter. The rhizomes had been coated by silica from a flood of hot-spring water, and the plant probably killed by the event, but since the air could get at the remains decay had set in and the plant had been destroyed rather than silicified.

'Well, I suppose this is encouraging in that we can see the same sort of features that exist on the sinter aprons deposited by the hot springs at Yellowstone, but we are not discovering how the Rhynie cherts that contain well-preserved plants were formed.'

The Rhynie stream area was very similar to Tangled Creek near White Dome geyser in Yellowstone. Tangled Creek has its origin in a continually erupting hot spring, where the water deposits silica as it cools and flows away from the vent as a stream. By some two kilometres from the vent the water has cooled to around 20°C, and provides a warm environment for plants and insects, but silica is still being deposited. Further downstream the water enters a marshy area, and this continuously wet area is possibly the type of environment where the Rhynie chert formed. Conditions would have been reducing where the plants were silicified to preserve the organic plant matter (Fig 2.7).

We were now closer to the geyser, and it blew again.

'That's 23 minutes since the last one,' announced Lyall. 'Prepare for a shower.'

We were directly down breeze, and the mist from the eruption shower was coming our way. The sunlight refracted through the water droplets produced a superb rainbow, and since Rhynie is a gold-bearing, hot-spring system there

Fig 2.7 A marsh on Tangled Creek in Yellowstone National Park fed by hot springs. The trees have been killed by hot water and have 'white socks' of silica deposited by evaporation. Many plants grow and thrive in the warm (*c*.20°C) water. The Rhynie chert was probably formed in a similar marshy environment.

probably was a very little gold at the end of the rainbow. Lyall was busy recording the time and duration of the eruption; it was easy to tell what he was thinking—would he be able to record the oldest-known, regular geyser activity and put it on the geyser-gazers web page. That should blow the minds of the geyser-gazer anorak fraternity. After a brief geyser shower we soon dried off, but a fine white deposit remained. The cooling spray had gently coated us with silica.

We were now picking our way carefully, and somewhat hesitantly over hot ground. Steam rose from some areas that were white or grey and devoid of visible life. There was a rotten eggs smell of H_2S in the air, and boiling water bubbled urgently from small vents. Cooler areas ranged in colour from bright brick red to green. These were coloured bacterial mats, and at least indicated temperatures below 70°C. We both felt guilty at walking on the mats; it was a cardinal sin to mark the mat surfaces in Yellowstone, although buffalo footprints and dung piles were always in evidence. Here we were, leaving great big boot-prints and hoping they would not be preserved in the geological record.

Vents were all around us. Some were quiet but with steam rising and crystal clear water revealing bright blue and orange colours, while others were small geyser vents with typical knobbly sinter shapes in the splash zone around the vent (Fig 2.8). Identical sinter textures have been found in a chert block from Rhynie, proving that there were geysers at Rhynie. The land surface here was generally hard, but we both knew that crusts could be thin. Progress was slow, and several times our feet sunk several centimetres into hot mud. This was clearly a very active area. There were patches of *Rhynia* and *Asteroxylon* that had been killed by the heat from below, that is a sure indicator that the ground temperature is increasing. Exactly the same feature can be seen in Yellowstone where new hot areas kill pine trees that become brown and withered; the trees die because the roots are burnt by the heat.

'We must get out of here—this area should have US National Park boardwalks with handrails, and we don't have a ranger in uniform to make us feel safer.'

'I would certainly be happier following Smokey through here,' Lyall agreed, with a nervous laugh.

It took us a while to pick our way through the hot stuff, past several geyser vents which were clearly active but on a longer timescale than our regular spouter. Then it was time to leave and head back to the Bus. We had been engrossed in plants, beasties and springs for the past six hours, and it was going to be dark in two hours' time. Rather than traipse over the hot springs area and risk putting a foot into something too hot, we decided to follow the outflow terraces of the geyser mound

Fig 2.8 Modern, hot-spring vent in Biscuit Basin, Yellowstone National Park (Shell Spring). The rounded sinter deposits typically form in the splash zone around the vent. Ancient examples of this sinter texture have been found at Rhynie, providing evidence for ancient geysers.

down to the river and make our way back up stream to the Bus. This had the added science target of seeing how much of the hot-spring outflow was reaching the river, and at what temperature. In fact, with a few temperature measurements we should be able to do some back-of-the-beermat calculations on heat input to the river and see how they tallied with the data from the heat sensor.

Walking down from the geyser flat was uncommonly like Yellowstone without the lodgepole pines. We were picking our way over the sinter, between *Horneophyton* rather than sedges, and we were heading for the Old Red River rather than the Firehole River. OK, there are other differences, like no elk, bison, fishermen or tourists, but the merging of the sinter apron from the hot springs with the alluvial flats of the river was unmistakably familiar.

Our Very Old Faithful geyser erupted again. It really was a very regular one, and had been performing every 22–28 minutes ever since we had first been disturbed by the distant roar. We arrived at the river on a bend with the eroding bank on our side of the river. This was good for science, but bad for crossing the river. We were standing at the top of a four-metre high, crumbly cliff and looking down into fast-flowing water—far too deep to wade. The river was eroding the outflow sinters and provided an excellent section of the sinter pile, if only we could look at it. Apart from getting safely down the cliff, there was nowhere to stand at the bottom. Deep water flowed swiftly at the base of the cliff. We would be better off standing on the opposite bank and viewing from afar. Water was flowing over the edge of the cliff into the river, and also issuing as underground streams from features in the cliff.

We had a choice: upstream towards the Bus and over all the water, crossing hot ground and outflow streams; or downstream away from the Bus, and back up the other side where there did not appear to be any hot springs. The question was how far would we have to go before we could cross safely?

Lyall did not like hot ground without boardwalks, so it was safety first and the long walk home. About a hundred metres downstream the cliff died out and the river became shallower but rocky; the sinter buildup had interrupted the river profile, and the gradient increased suddenly below the hot spring area. Crossing here was going to be a rock-hopping exercise, but the 'rocks' were no more than eroded blocks of sinter terrace. We could now see the river for hundreds of metres downstream, bubbling and rushing on its now steeper course into the Orcadian Basin.

We decided to cross here as the lesser of two evils. The river bed was firm in places, cemented by silica, and bubbles rose from between pebbles in the water. There was clearly hot-spring activity beneath. We took temperature readings and soon found that water at 70°C was entering the river through the river bed. No wonder the river was significantly hotter downstream. In retrospect we were lucky; we both lost our footing in a deep fast channel and floated off downstream, arms outstretched, and knees up. The action known to salmon fishermen as the Hugh Falkus manoeuvre. Good for Hugh; two more lives saved. We emerged bedraggled and with water in our boots, but none the worse for the experience, and we had crossed the river.

Sitting on the bank we reviewed the situation. We were now well behind time, and had to leg it as quickly as possible to the Bus. At least we were not cold; the stream temperature was 22°C—there must have been a lot of hot water coming into the river from the geyser area. Setting off upstream, and following the bank as closely as possible, we were soon back at the sinter cliff (Fig 2.9). Looking across the river the layering of the eroded sinter terrace was clear, and there was one major irregularity where an old hot spring vent had been filled by later sinter. It looked like a section of an inverted cone and the harder sinter that had formed the vent walls stood out from the cliff. There was only time to remember the features—we were not unpacking equipment again. The geyser, now firmly

Fig 2.9 Sketch of the riverside cliff eroded in hot-spring sinters deposited by geyser activity. A cross-section of an earlier hot-spring vent is seen in the cliff.

christened 'Very Old Faithful', was nearing eruption, and this time the geyser spouted for a full two minutes and a wave of hot water swept down the shallow channels towards the river, where it cascaded over the sinter cliff in a shower of steaming hot water. Another twenty-five minutes had passed and we were now only back to the point where we had made the initial decision to go downstream rather than upstream over the hot spring area.

Accompanied by sloshing noises from Lyall's boots, we plodded on upstream. To our left a dark, rounded hill of gravelly looking material was clearly different from other hills in the area. Its sides were gullied and eroded, and it was virtually bare of vegetation of any kind. There was no time to divert to climb it, but there was evidence of its nature in the debris being washed down rivulets to the main stream—lots of black to grey, irregularly shaped fragments of rock with numerous air bubbles and glistening, glassy, tabular crystals. This was volcanic ash, and our hill had been a small volcanic vent, but was now inactive and undergoing erosion. When the river flooded, this material would be spread far and wide, and might even be washed into the lower sinter surfaces.

We had hardly paused to look at the volcanic rocks; we were pushing on for the Bus. The sun was nearly on the horizon, and at this latitude it would be dark half an hour after sunset. Long shadows were creeping over the land, and we walked as fast as possible, picking our way over sandy and muddy surfaces and skirting any wet, muddy areas that might be soft under foot. There was a faint unpleasant smell of rotten eggs in the air for which neither of us claimed responsibility. The next mud pool provided the answer: it was gently steaming and the mud periodically erupted as bubbles of gas were forced to the surface.

'Still a bit hot here then.'

'Yes, but we must leave it.'

It was the backward glance that brought about panic. Lyall was right beside me when he suddenly lurched to one side as a foot disappeared down a hole. He was on one knee but the other leg was straight but buried.

'Quick! Pull me out—it's hot.'

Grabbing him, I heaved, and Lyall's disgustingly muddy and smelly foot emerged from a tiny mud vent barely big enough for his foot to fit inside. The mud was hot and clung to his boot. Scraping it off with my gloved hand I could feel the heat.

'Are you burnt?'

'It's certainly hot, but my foot seems OK.'

It seemed that the water sloshing in his boots had actually been of benefit in preventing bad burning. We had no option but to continue with caution. It was nearly dark when we crossed the rocky ridge and the Bus came into view; soon we were there, and let ourselves into the entry port. We had seen many things and confirmed that Devonian hot-spring areas were just as hazardous as Yellowstone today. We watched the sun set behind an ancient Tap o' Noth, and then started the Bus and headed homewards.

Excursion 3

Fishing in Caithness

TIME: Mid Devonian.

LOCATION: Caithness, near Red Point.

OBJECTIVES: To examine the shores of the ancient Orcadian lake, and do a little fishing, on the pretext of studying the fish fauna.

THE MODERN EVIDENCE: The lake deposits of the Middle Old Red Sandstone of Caithness and Orkney, and the superb fish fossils found in the rocks of those areas.

Watten, Harray, Stenness and Broadhouse are names of lochs on the map of Caithness and Orkney. They have two factors in common. The first is that they are famous trout lochs—the object of pilgrimage by fly-fishermen—and the second is that they all lie on the rocks of the Devonian flagstones of this part of Scotland.

The flagstones are thin-bedded, grey, brown and green rocks which occur in layers with various thicknesses and strengths suitable for paving flags or roofing tiles, and are superb for building field dykes and habitations. In Caithness large flags are set on edge to make distinctive field boundaries. The value and use of the flags has been recognised from the earliest times. The ancient village of Skara Brae on Orkney, the Iron Age stone towers known as brochs and the great standing stones of Stenness in Orkney are all constructed of flagstones. More recently, city centres throughout the world have been paved with Caithness flags. After a decline in the industry in the earliest part of the twentieth century, this marvellous material is enjoying a comeback at the start of a new millennium, as architects at last choose natural materials in place of concrete slabs.

The quarrying activities in Caithness and Orkney frequently expose dark grey to black flagstones that contain beautifully preserved fossil fish. Famous localities are Cruaday Hill in Orkney and Achanarras Quarry in Caithness (Fig 3.1). From these quarries thousands of specimens have been found that form the subject of many scientific works describing their zoology and evolutionary importance.

Fig 3.1 Flagstone tips at Achanarras Quarry, near Spittal, Caithness. Fifteen different genera of mid Devonian fish have been found at this world-famous locality in the Achanarras fish bed. It is a protected site (SSSI), but collectors are allowed to search the quarry tips for fossils.

Fossils from these, and many other localities, have found their way into museum collections worldwide and provide a snapshot of a fish fauna inhabiting the area some 380 million years ago.

The flagstones are the deposits of an ancient lake, dubbed Lake Orcadie. At its greatest extent the southern shore of the lake lay south of the Moray Firth from Gamrie to Fochabers, Nairn and Inverness, and stretched northwards to Caithness and Orkney, and on to Shetland.

Scotland lay about 20 degrees south of the equator at this time, and climatic variations between hotter and dryer, or cooler and wetter, conditions caused the lake to expand and contract in time with climatic changes that affected runoff into the lake and evaporation. Consequently, the deposits of the lake floor record conditions ranging from deep lake to an exposed playa lake bed with large polygonal sun-cracked surfaces, a feature seen in miniature in any sun-dried muddy puddle.

The same variation in lake size with climate occurs in continental lakes today. Lake Eyre in Australia, now little more than a large salt pan, has expanded to three times its present size with waters seventeen metres deep in the recent geological past. The climatic changes that brought ice ages to Europe resulted in rainy and cooler periods in Australia, and lakes expanded. Thus the lake deposits provide a record of lake expansion and contraction, and indirectly of climatic variation.

The fossil fish of Caithness are found in dark grey to black flagstones which are very finely laminated or layered. Under a microscope a thin rock section shows three types of laminae are present, and these occur in a regular order. The scale is so fine that all three occur within a millimetre.

The laminae are:

> Organic—black organic carbon;
> Carbonate—usually calcium carbonate, the mineral calcite;
> Clastic—a fine sand of quartz, feldspar and mica mineral grains.

The interpretation of this depositional triplet is that it represents deep lake deposits in a seasonal climate. The clastic laminae were deposited in the rainy season when rivers brought silt into the lake. The carbonate laminae formed in summer when water temperature increased and carbonate was precipitated due to the activity of photosynthetic algae; and the organic laminae are the remains of the organic sludge produced by autumnal death and decay of the algae. This is a bit of a generalisation since there were several ways in which the clastic laminae could be introduced to the lake, one of which was fallout from dust storms blowing over the lake.

In some fish beds the thickness of these repeated laminites indicates that they were deposited over long periods: the fish bed at Achanarras Quarry represents about 4,000 years of Earth history. There are subtle variations in the thicknesses of the laminae within the fish bed, and following detailed measurements Stephen Andrews, in his research work at Aberdeen University, has shown that the changes in thickness have the same periodicity as modern sun-spot cycles. Sun-spot activity exerts a short-term control on climate today and seems to have played the same role in the Devonian. The fish fossils at Achanarras represent changing fish faunas spanning about the same breadth of time that separates ourselves from the peoples who used the flagstones to build Skara Brae, the famous ancient village preserved at the Bay of Skaill in Orkney.

How then did the fish come to be so well preserved? They are crushed flat, but in many cases every fin ray is in place, and there is no sign of damage to the fish. In other instances the fish carcasses clearly decomposed, and a scatter of shiny scales is all that remains. The preservation of organic carbon layers in the flagstones points to an environment in which oxygen was absent and organic carbon could accumulate. The perfect preservation of both the fish and the delicate lamination indicates a lack of burrowing animals and scavengers at the lake floor. If a fish dies (usually killed) in a pond or river it is eaten by the predator that killed it, or it is consumed by scavengers. Nothing usually remains to be fossilised. In the flagstone fish beds the fish present must have been killed, but not predated or scavenged.

The fish present as fossils at Achanarras include bottom-dwelling species such as *Pterichthyodes* (Fig 3.2), the 'winged fish' discovered by Hugh Miller the Cromarty stonemason and author of The Old Red Sandstone. This fish lived in areas where the lake bottom was oxygenated, and it could grub for morsels like a vacuum cleaner. So, how did it die and come to be preserved in the deep lake? There are several possibilities, and one or all may provide the answers.

Lakes that have cold anoxic bottom waters usually have warm oxygenated surface waters, and shallow margins where animal and plant life exists. However, if the balance of the stratified lake waters is upset, maybe by a large storm, the mixing of poisoned bottom waters with the surface water may reduce the oxygen content to such an extent that the fish die by asphyxiation. Since all die, there are no scavengers left to clear up the mess.

Another possibility is connected with 'algal blooms'; algae (and cyanobacteria) may proliferate to such an extent that they use up the available oxygen, and their decay effectively poisons the water. Algal blooms can be accentuated and triggered by man—as by nitrate fertilisers washed into lakes and rivers—but are also a common natural phenomenon.

A third idea is that as lakes contract and evaporate, salt concentrations rise and eventually exceed the tolerance of the fish. An incident of this type took place in Australia in 1976. Exceptional rains caused by monsoonal conditions spilling

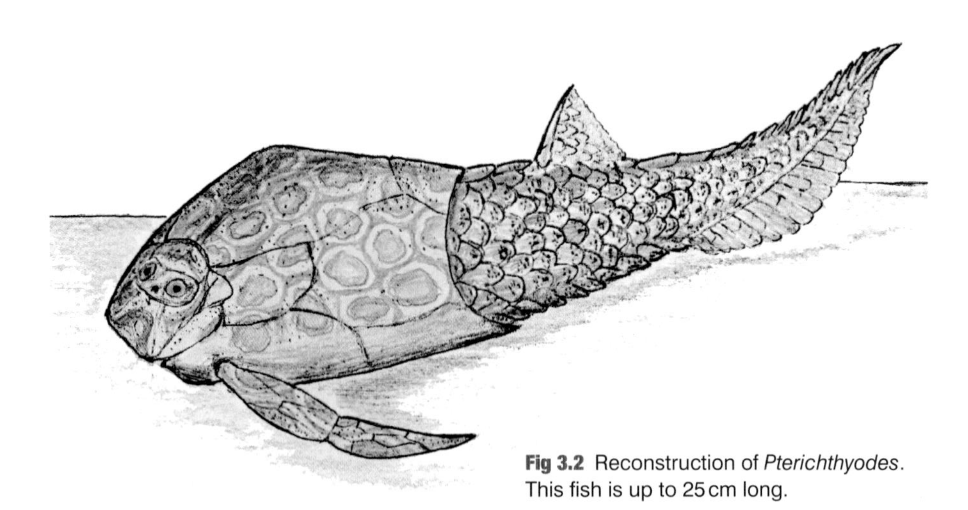

Fig 3.2 Reconstruction of *Pterichthyodes*. This fish is up to 25 cm long.

over the Great Dividing Range brought flooding to the salt flats of Lake Eyre, the water depth rose to eight metres and the freshwater flood brought with it fish. The fish flourished, and birds then arrived in the desert for the unusual feast. As the waters retreated over eighteen months, conditions became saltier and the fish died and dried in their thousands along the lake side, producing an odour effective for more than fifty kilometres.

Yet another possibility for the death and preservation of fish is water temperature. In very hot weather the temperature of streams or shallow lake waters can rise above the tolerance level of particular fish and they are killed. The extreme heat in the New York area in early July 1999—more than 38°C for several days—caused mass mortalities of trout and suckers when the water temperature in rivers rose to more than 26°C. Similar fish kills owing to heat also occur in the Great Lakes of Canada.

Any or all of these mechanisms could have affected Lake Orcadie. In each case a mass mortality of fish resulted. The dead asphyxiated fish—they are often preserved with the spinal curvature typical of asphyxiation—started to decompose internally, the carcasses became bloated with gas and floated out into the lake. Eventually each carcass bursts, the gas escapes, and the fish sinks into the dark anoxic depths of the lake (Fig 3.3) where it can be gently covered with the silt and mud of successive seasons of lamination deposition. Burial, and consequent compression, of the sediment pile converted the muds and silts to the solid cemented flagstones seen in the quarries of Caithness today (Fig 3.4).

I have spoken only of fish; what else is found in the fish beds? The short answer is 'very little', but there are drifted fragments of plant stems and very rare remains of arthropods. There is nothing to suggest that the lake was directly connected to the sea. At the same time as the fish beds were being deposited in North Scotland there were coral reefs in the south of England at the edge of the Old Red Sandstone continent. These reefs contain a wide range of Devonian marine fossils—corals, trilobites, brachiopods and crinoids included. The lack of any of this fauna in the Orcadian lake, despite the abundance of carbonate, strongly suggests non-marine conditions in the lake.

The next obvious question to ask is 'how did the fish get into the lake?' Fish initially entered fresh water from the sea, probably in the Silurian Period. The Devonian fish found in Caithness and Orkney had close relations in areas that were widely separated in Devonian times (for example, China and Australia). Migration of fish must have been through the marine environment, but fish invaded rivers and lakes in the various continents of the Devonian. Thus it is

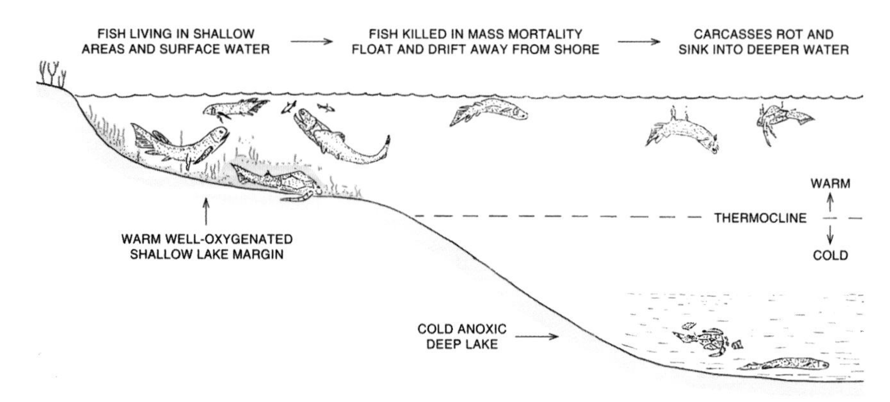

Fig 3.3 Cartoon sketch of the formation of the Achanarras fish bed. Fish living in shallow lake margins are killed during a mass mortality. The carcasses, bloated by gas, float out into the lake where they eventually sink to the floor of the lake in deep water. Since there is no oxygen in the deep water there are no scavengers to disturb the carcasses, and they are gradually covered by mud.

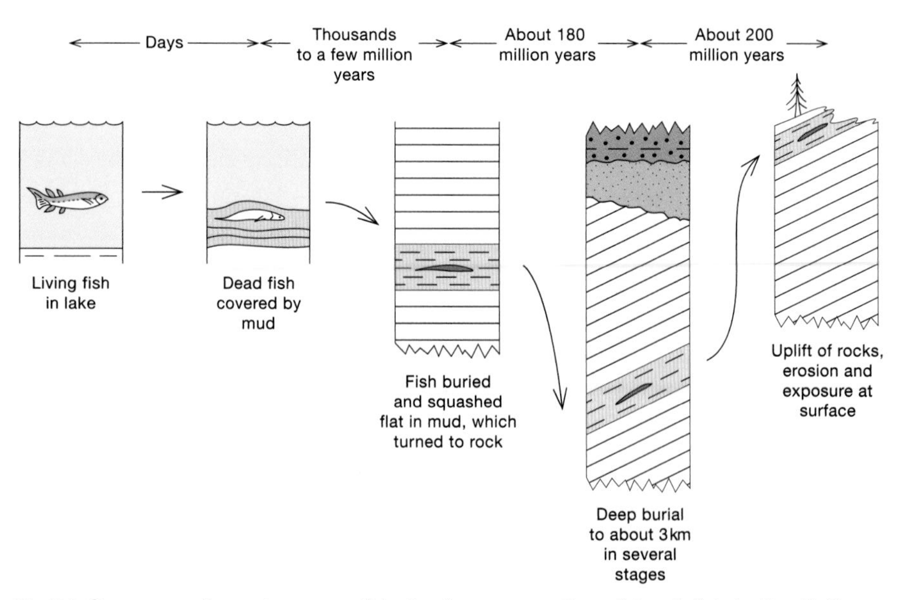

Fig 3.4 Sequence of events responsible for the preservation of fossil fish in the Caithness flagstones.

likely that at times of high lake level a river connected Lake Orcadie to the sea, and the fish migrated up the river to reach the lake (Fig 3.5).

Before going fossil fishing it is essential to know the quarry. A fish spotters' guide (Fig 3.6) to a few of the Devonian fish of the Orcadian lake is therefore supplied to complement the storyline that follows.

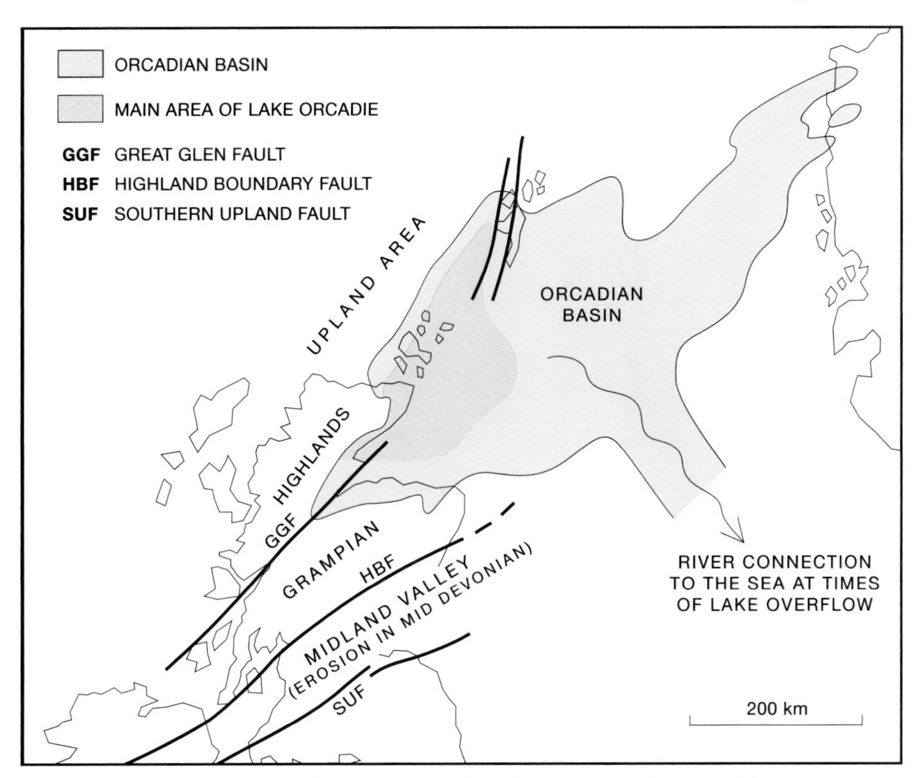

Fig 3.5 Palaeogeographic map of Lake Orcadie in the mid Devonian of Scotland.

At Achanarras Quarry in Caithness fishing ancient and modern meet. The waste tips of this disused quarry have produced some of the best-quality mid Devonian fish fossils from anywhere in the world, and can be 'caught' with hammer and chisel. The waters of the small lake that partly fills the quarry contain thin ravenous little trout, introduced some years ago, and surviving on the pond life that has colonised the quarry. Almost any kind of fly will serve to catch these lean and hungry fish. A few casts can produce a trout that can be laid beside its most ancient ancestor, *Cheirolepis*, on the shores of the quarry lake. Such coincidence makes Achanarras an ideal starting point for our fishing excursion to the Devonian Lake Orcadie.

Jeremy, a long-time academic colleague and fishing companion, had agreed to join me on the excursion. We had fished together in modern Scotland and Australia, and ancient Caithness seemed an exciting prospect. We initially considered a fish-watching submersible excursion, but the research grant would not stretch to cover the extra expense, and it could not be justified as essential to the scientific objectives of the work related to climate change.

Fig. 3.6 Fish spotters' guide. Specimens and reconstructions of some of the fish of Lake Orcadie.

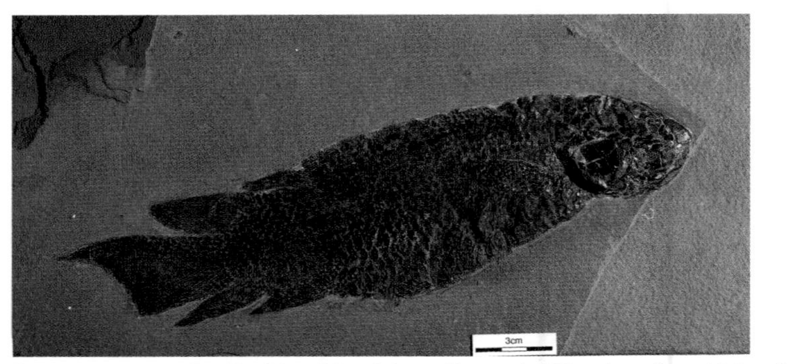

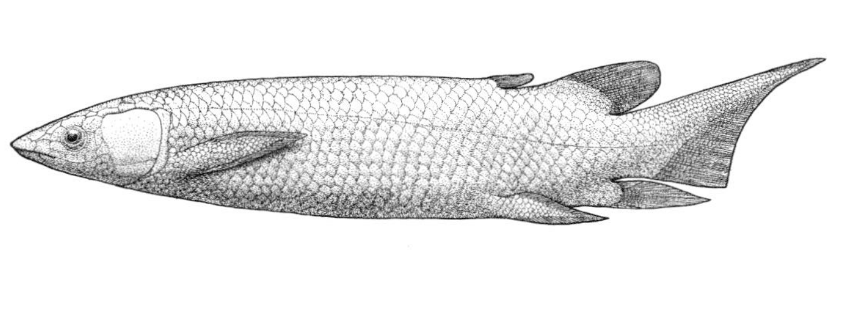

Name: *Dipterus* **Family:** A lungfish, relatives still occur in Africa, Australia and S America. **Size range:** Up to 40 cm but few bigger than 25 cm. **Diet:** Omnivore. Had palatal tooth plates. Plants, invertebrates. **Habitat:** Survived low oxygen levels, shallow muddy areas.

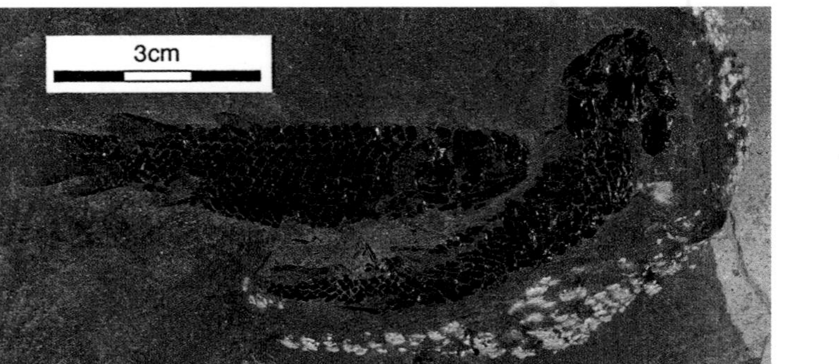

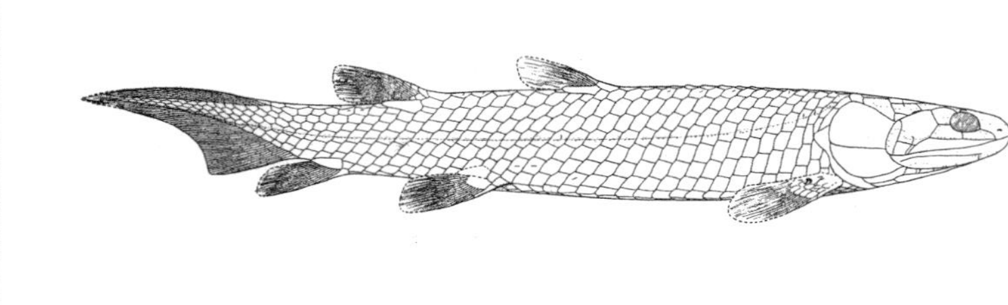

Name: *Osteolepis* **Family:** A Crossopterygian, or 'lobe-finned' fish, a distant relation of the Coelacanth. Several species in the Orcadian lake. **Size range:** Most 10–20 cm, but a few reached 30 cm. **Diet:** Small sharp teeth. Fed on fish and arthropods. **Habitat:** Round-bodied, mid-water swimmer, possibly a shoal fish.

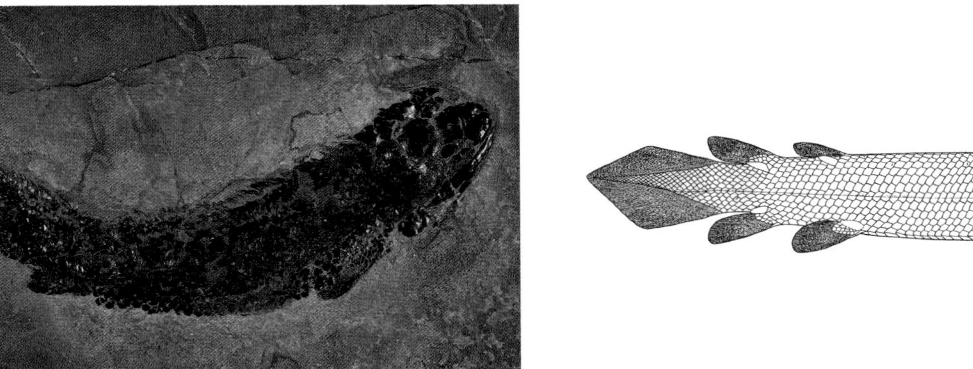

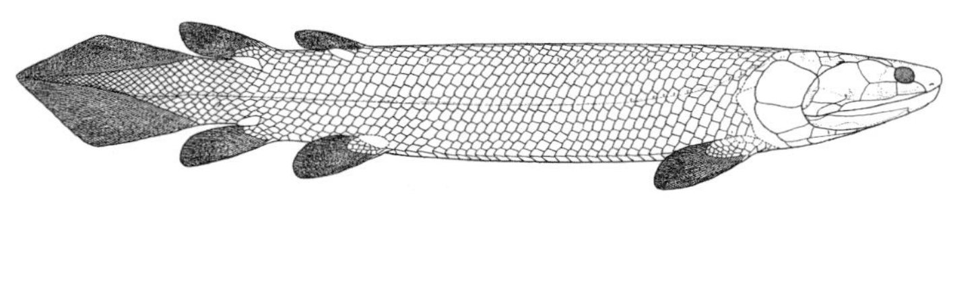

Name: *Gyroptychius* **Family:** A Crossopterygian, lobe-finned fish. **Size range:** Most found are 25–35 cm long. **Diet:** Sharp teeth, mainly a fish-eater. **Habitat:** Lurking predator. Powerful tail for fast acceleration.

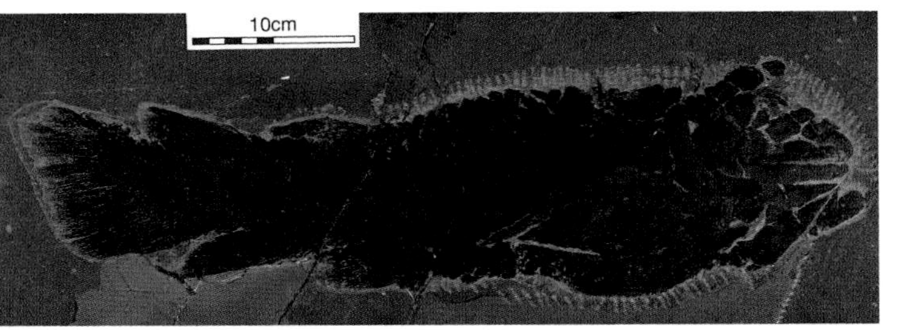

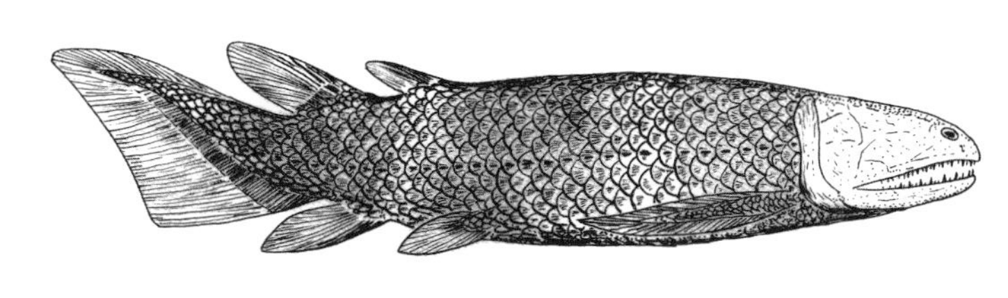

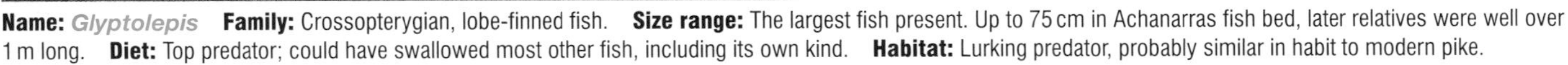

Name: *Glyptolepis* **Family:** Crossopterygian, lobe-finned fish. **Size range:** The largest fish present. Up to 75 cm in Achanarras fish bed, later relatives were well over 1 m long. **Diet:** Top predator; could have swallowed most other fish, including its own kind. **Habitat:** Lurking predator, probably similar in habit to modern pike.

Fig. 3.6 Fish spotters' guide (*continued*).

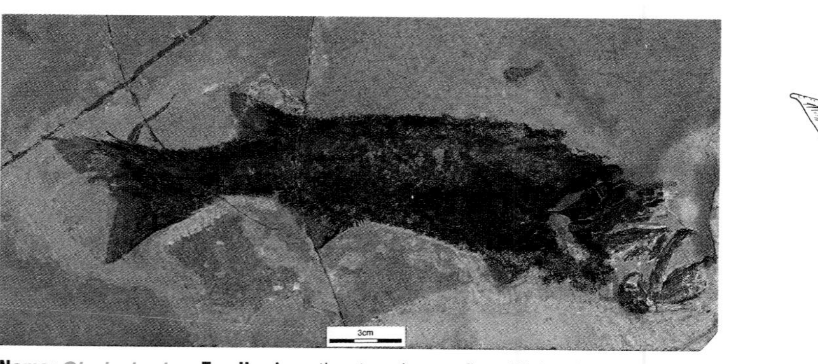

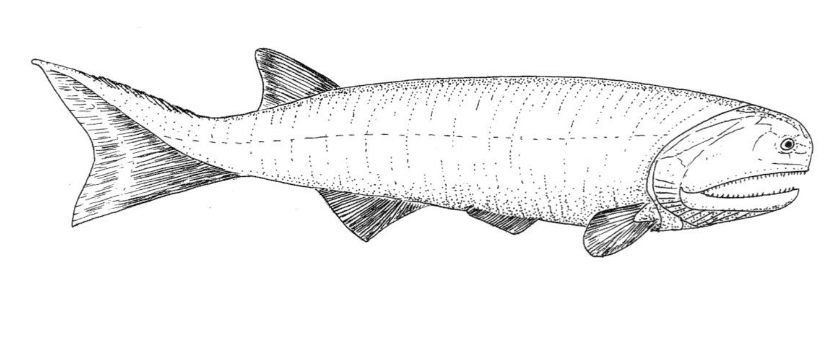

Name: *Cheirolepis* **Family:** An actinopterygian ray-finned fish, related to modern bony fish. **Size range:** Up to 40 cm. Most specimens found are 20–30 cm.
Diet: Small sharp teeth suitable for eating fish, arthropods. **Habitat:** A round-bodied fish, hunting in mid water.

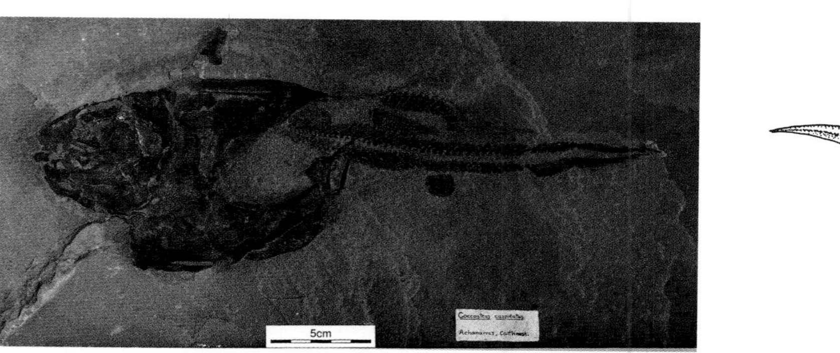

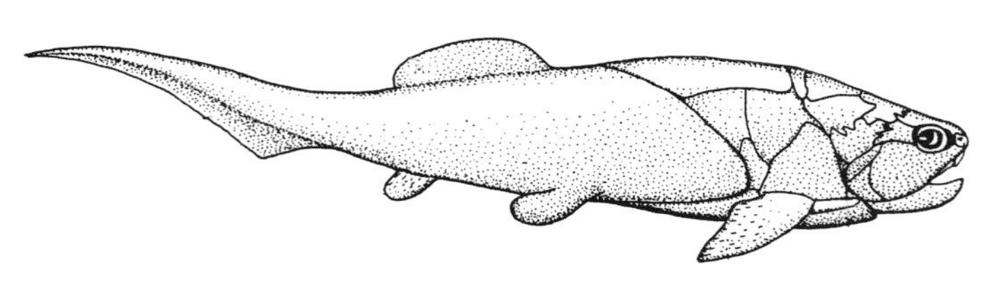

Name: *Coccosteus* **Family:** An armoured Arthrodire. Extinct group of Placoderm fish. **Size range:** Most specimens about 30 cm, juveniles rarely found.
Diet: Carnivore, Dipterus and acanthodian bones found in stomach. **Habitat:** Probably an ambush predator.

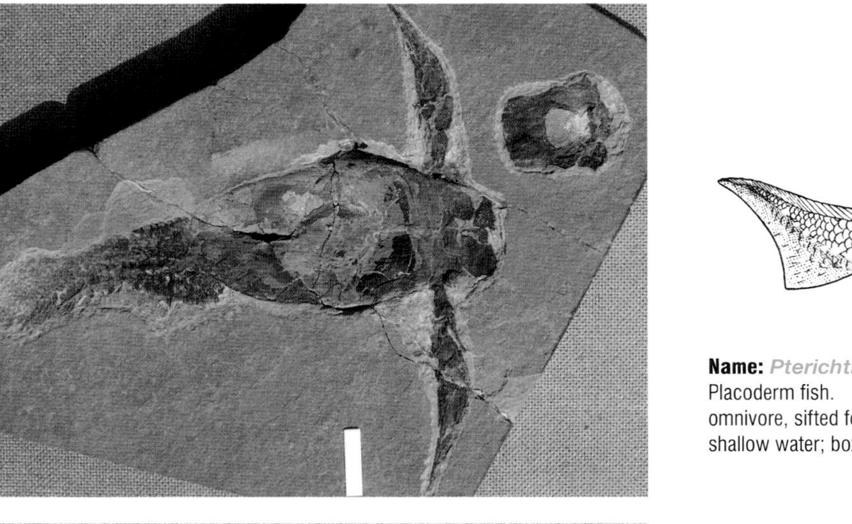

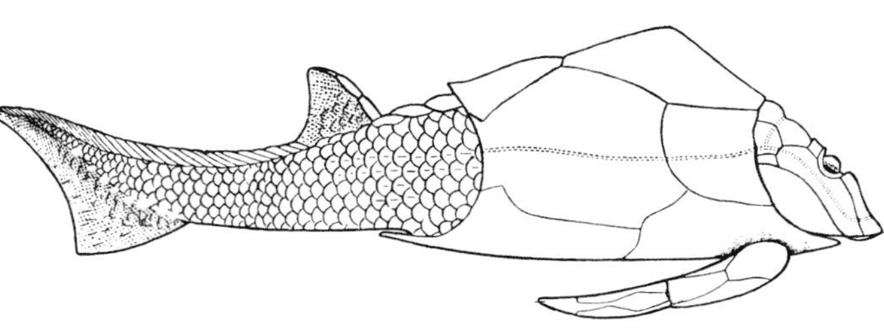

Name: *Pterichthyodes* **Family:** An armoured Antiarch. Extinct group of Placoderm fish. **Size range:** All sizes from 1 to 35 cm. **Diet:** Weak-jawed omnivore, sifted food items from mud and sand. **Habitat:** Bottom dweller in shallow water; box armour for protection.

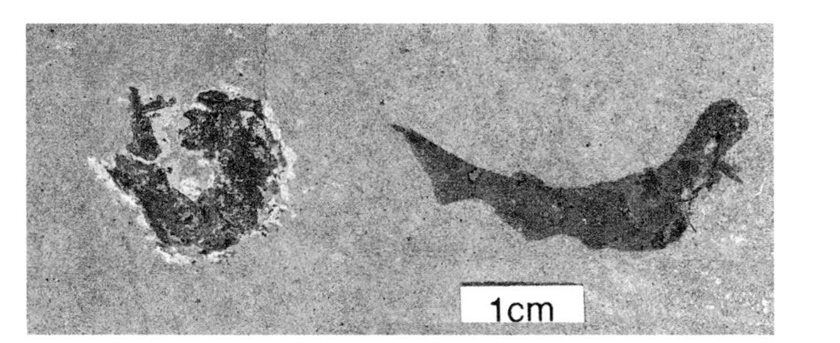

Name: *Mesacanthus* **Family:** Acanthodian fish, spines support fins. **Size range:** Mesacanthus under 5 cm, but some other genera of acanthodians up to 30 cm long. **Diet:** Toothless. Fed on small items, micro-organisms and vegetable matter. **Habitat:** Probably a shoal fish, sometimes very abundant.

If we wanted to study a fish and its lifestyle we would have to hold the fish and examine its gut contents to find out what it eats. We therefore decided that we could learn a lot, and be well entertained, by fishing along the shoreline of Lake Orcadie. Also, from our fossil studies we had concluded that most of the fish lived in the shallows at the lake margin, in water too shallow for a manned submersible to operate. Given our mutual interest in fishing, the outcome was inevitable.

So we set out on a mission to the Devonian carrying a compact variety of fishing tackle, including some innovative 'flies' (no natural bait allowed, the Devonian was strictly 'fly only'). A broader remit for the trip was to study the response of the lake to climate change. This was the main objective of the excursion, part of general scientific interest in how our planet and its biota responded to climate change in the past, and what the human race might expect in the future. We planned to watch and time contractions and expansions of the lake in a scan of 10 million years of lake history. We would do this from orbit before landing and during the process pick a time and place for our landing.

Our plan was to land on the shore of the lake at a time of high lake level; we had no interest in seeing a dried-out, lifeless playa surface. Given that the lake expansions and contractions were geologically rapid, the timing had to be precise to a few thousand years. As the time log tracked back we 'parked' the Bus in our programmed orbital position about over the present location of Thurso. This is of course not strictly a stationary orbit position, because in the Devonian Scotland was drifting north towards the equator.

As the time-tracking factor slowed on reaching 380 million years, and we had our first chance to view the earth below, a magnificent scene unfolded. We were at a margin between a flat brown plain with scattered hills that seemed to form islands and an upland area with mountainous peaks, some clearly volcanoes. The land was shades of brown, yellow and ochre with small patches of green nestling in valleys draining the high land. We had clearly stopped the clock at a dry time in the history of the lake. We now needed to pinpoint the maximum extension of the lake; the time the fish were being fossilised at Achanarras.

'I think we are likely to be near the end of the lake's existence,' I pointed out to Jeremy. 'If we find the end point we can track back and record the whole lake history. If we scan at 10,000 years a minute we should be able to see the lake expand and contract, and eventually die. The periods we expect are of the order of 20,000, 40,000 and 100,000 years, and so we should have some results within the hour.'

As time steadily changed there was no great alteration in the view for ten minutes. Then, very suddenly, within less than a minute, an amazing transformation occurred. Blue waters spread over the brown lake floor, and simultaneously a greening of the land took place with green tinges extending ever farther up hillsides, mountain lakes expanding and bright rivulets of water flowing down valleys to the lake. The view was not as clear as before, probably caused by fuzziness because of the greater cloud cover, but the main message was clear—a sudden climatic change from arid conditions to an extensive lake with vegetated shores. In another two minutes the scene had reverted to one much the same as our original 'brown' view. Several more transformations took place as we watched, and the brown period extended in time and the blue-plus-green transformations became less extensive and less frequent; the lake seemed to be filling with sediment and drying out.

'OK, let's stop here and work our way back,' I suggested.

And so we changed time-travel direction and proceeded back in time at 10,000 years per minute. With our fixed viewpoint from the Bus, high above the lake margin, it was a simple matter to film and record the timing and extent of the brown and green-plus-blue scenes. It was this information that would confirm or refute earlier suggestions that the lake was under a celestial control, responding to periodicities of the orbit of the Earth and Moon system around the Sun. I could be proved wrong with my previous estimates on timing, but that matters for little; science must progress, and this technology provided answers to questions with which we had been struggling for years.

Some ten hours of recording later, representing six million years, Jeremy and I found our target. There had been false hopes on several occasions, but suddenly we saw the mother-of-all-transgressions of the lake. The water extended up valleys and lapped against hillsides not reached by any previous lake phases, and what is more it stayed high with only minor fluctuations for fully thirty seconds—5,000 years' worth of time.

Moving to a lower altitude of about 10,000 metres we could make out details of the lake shoreline. We were searching for our target spot, coinciding with present-day Red Point to the west of Dounreay in Caithness. Our choice was determined by the fact that at that locality we could stand on the ancient shoreline, at the time of the existence of the fish found at Achanarras. We needed a lake margin, or even river locations, where waters were oxygenated and the fish had their living area. It was no use dropping in at Achanarras, where the water was deep and the bottom conditions anoxic. Red Point, however, included shoreline where the lake lapped against bare rock, and an area where there was a beach and sandy sediment. Also, the whole escapade as far as the fishing went could be a total disaster. We could not pick our day or choose the weather. In true fishing tradition it could be too hot, too cold, too bright, too calm or too windy. The potential excuses were endless.

The lake margin was clearly complex in the chosen area. We could see hills and drowned valleys—the fingers of blue lake waters extending kilometres up the valleys and ending with a small delta or sand flat where each stream entered the lake. There were no major rivers to be seen, however. Hilly areas formed headlands, and there were attractive rocky islands protruding from the lake waters.

Decisions on a landing spot had to be made quickly, and we picked an interesting hilly rocky headland that was adjacent to a bay with beach ridges. Hopefully we would have a look at a steep and gently sloping lake margin. We would also be able to see what vegetation and animal life existed on eroding rocky surfaces as well as on sandy lake shores. There appeared to be a variety of both terrestrial and shallow water environments within a short walk.

Co-ordinates were fixed, and we took the Bus down to the Devonian surface. The landing spot was ideal: a flat sandy area in a valley a few hundred metres from the lake shore. A waterless channel ran down to the shore, and on either side rocky hillsides rose above the Bus. We had eight hours for our project, and we quickly gathered our gear and stepped out of the Bus and set off for a lakeside stroll (Fig 3.7).

Fig 3.7 Route of Excursion 3 on the shores of Lake Orcadie.

It was a relief to leave the Bus. The exchange of cramped claustrophobia and the myriad of flickering screens and instrument displays for the silence of a fine Devonian day was invigorating to say the least.

Jeremy and I set off westwards up the valley side picking our way over granite and gneiss boulders in various states of disintegration. Some were hard and fresh; others were crumbling beneath our feet into grains of quartz and feldspar—the source of the Devonian gravels that fringe Red Point at the present day. Patches of lichen dotted the rocks—no real difference from earlier observations at Rhynie during Excursion 2. Small plants with leafless stems struggled for existence between the boulders, and undoubtedly if we had stopped to look we could have found beasties similar to those found in the Devonian Rhynie chert in Aberdeenshire; but we were after larger prey.

From the top of the hill we could look back east to the Bus, neatly parked in the valley. To the north the lake shimmered in the morning sun. Only small ripples washed the rocky shore that reflected shining white in contrast to the pink and grey hues of the granite we were standing on. A rounded island of granite poked through the waters about 200 metres offshore, clearly a continuation of the headland on which we stood. We could well have been looking at Red Point in that island; after all, it sticks up through Devonian sediments in the present-day outcrop and must have been a hill drowned first by the waters and then by the sediments of the Orcadian Basin.

To the west the ground fell away to the next valley, where a little water remained in a pool where the sandy river bed was blocked by the beach ridges at the lake margin. Maybe this was the dry season, but the river had no open outlet to the lake; maybe a flood would break through the ridges. The ridges looked interesting in that they were at different levels and there was clear evidence that the lake waters had been higher. We were not at the time of maximum transgression, but at least we were close. Jeremy and I took in the scene, minds divided between scientific observation and evaluation of the fishing prospects.

'Let's go north along the ridge to the lake margin,' I suggested. 'I want to know why the rocky shore is white, and we need to look for freshwater plants and animals on a rocky lake shore. Then we can go west into the bay where there is a sandy shore with beach ridges and shallow water.'

'Good idea,' agreed Jeremy. 'It looks quite fishy out by the point and going round into the bay. I would fish there if it were an estuary in Western Australia.'

We (more Jeremy because he is originally Welsh) chatted in fishy anticipation as we descended the ridge to the shore. Approaching it, the white strandline demanded our attention, and speculation started.

'Evaporites? Salt or gypsum?'

'Carbonate? Stromatolites, with aragonite in splash zone as at Shark Bay in Western Australia,' countered Jeremy.

Speed of progress overtook the speculation, and revealed a favourable composite answer. Gypsum was present at high levels on the beach and crystallised in splash pools in the rocks, but most of the white was carbonate, generally porous and fine-grained and forming a rim at the water level. Beyond the rim the water was clear, and carbonate coated the rocks under water and grew into bulbous forms on the tops of submerged pebbles and boulders. These were algal stromatolites, known as fossils from similar situations in the Orcadian Basin. In the modern world they are best seen as fossils in the Stromness flagstones on the cliff top at Yesnaby on mainland Orkney, but can also be found on the shore at Stromness and near Dirlot Castle on the Thurso river (Fig 3.8). A good place to see living stromatolites is in the Shark Bay area of Western Australia, where a variety of forms grow along the shore of Hamelin Pool in water of about twice normal marine salinity.

Areas in the water not covered in stromatolite sported green filamentous growths looking like clumps of blanket weed, the scourge of pond owners. There were also tufts of a delicate plant with whorls of branchlets, resembling charophytes such as the modern stonewort, but we could not find the characteristic little spirally ornamented reproductive structures called oogonia. It was probably just the wrong season, because we know they had developed in the early Devonian, having been found recently in the Rhynie chert.

Offshore the water was dark blue, reflecting the sky, and the lake bottom shelved rapidly away. This was clearly a hillside drowned by the rising lake, but the lake had been stable for long enough for waves to cut a notch in the rocks at the shoreline and for stromatolites to grow in the shallow water. If the water retreated it was easy to see how a distinct white bench with beach deposits would be left cut in the hillside, just as is seen in the modern world on the hillsides bordering the Great Salt Lake in Utah and also bordering the salars of the antiplano in the Andes.

'Look! Fish,' cried Jeremy.

Sure enough, cruising gently past a few metres from the shore was a fish, and a fish of no mean size. All we could really make out was a large tail which combined

Fig 3.8 Stromatolites growing on a large pebble of metamorphic schist that was once close to the shoreline of Lake Orcadie. Algal stromatolites grew on the surface of the pebble while it was in shallow water. This specimen was found by Stephen Andrews at Dirlot Castle, Caithness close to the unconformity between Moine schists and Old Red Sandstone lake deposits.

with dorsal and ventral fins at the rear of the body to give the animal a powerful propulsion machine. The pectoral fins appeared to be acting as stabilisers, and the head was large and broad.

'How big do you think it is?'

It is always difficult to estimate size under water as any fisherman will know, having cast to a good-sized trout and caught it, only to find it was much smaller than expected (or hoped for).

'It must be a foot long, allowing for the magnification of the water, so it could be a large *Dipterus* or a *Glyptolepis*. I don't think it's an osteolepid; the head looks too broad and flat.'

As we watched, the fish made a speedy lunge towards us, and some small fish that we had not previously noticed scattered in all directions. The large fish gulped several times and resumed its patrol and was quickly lost from sight.

'Well, I reckon that was a *Glyptolepis* and he was attacking a small shoal of something,' Jeremy remarked.

We sat at the water's edge peering into the lake. Soon we made out movement as small fish emerged from hiding places between rocks. They were all small—no more than two centimetres long, and each had tiny spines supporting its dorsal and pectoral fins.

'It's a Devonian one-spined stickleback,' suggested Jeremy.

'Good idea, but it's more likely to be *Mesacanthus*.'

'So what do we have in the fishing box that looks like a *Mesacanthus*?'

Fishing was now taking over and the rod was extracted from its tube and a reel appeared from a backpack pocket, together with some leader material.

'*Glyptolepis* has good teeth so I reckon we will need to use the tough stuff,' I proposed.

This strong but supple material had superseded wire traces and could be used for fly fishing for very toothy critters.

The urge to catch the first fish we saw was overtaken by a bit of more controlled thought. We started looking for the small life forms that fish might eat. This involved searching under stones in the shallows and sifting through the sediment with a sieve. We did find some animals, but it was hard work. There were a few small clam-shrimps which ploughed along the surface leaving bilobed trails in the finer sediment patches and also small multi-segmented arthropods which were unlike anything known in the fossil record. Some of these were living in small burrows in the sediment and were reminiscent of the lifestyle of small *Corophium* shrimps that live in U-shaped burrows in modern estuarine mud.

When one large rock was lifted from the water, a flimsy brown object drifted out: it was an arthropod with a broad headshield and an abdomen of several segments; legs waved limply from its undersurface as it swirled in the eddy created by the lifting of the underwater rock. Now the pond net came into its own and the beastie was caught remarkably easily from where it had gently come to rest on a patch of gravel. It was immediately obvious that our beastie was not alive—but what we had caught was not a dead animal but the moulted exoskeleton of the animal. Like all arthropods, such as crabs or shrimps, it had to moult and had probably done so under the protection of the rock. In air the delicate moulted shell became limp and collapsed, the various parts disarticulating easily. It was obvious why fossils of such animals were so rare, yet something similar had been found at Achanarras, a shadowy impression of a chasmataspid—a very obscure group of arthropods.

Back to fishing. No time for obscure arthropods.

'What flies shall we try?' I asked.

After a short discussion, we concluded that something small and shrimp-like would be a good bet for any omnivores seeking small items of prey. We would also need something flashy as a small fish imitation to try and catch the big fellow we had already seen. The small arthropods we had caught seemed to live on or close to the bottom, so we chose a slow sinking line with a long leader and a slightly weighted, shrimp-pattern fly on a number ten hook.

Jeremy was the first to cast a fly into the Orcadian lake. The line shot from the rod tip and snaked across the water, rested and slowly sank. The automatic reaction was to fish as if trying to catch a rainbow trout in a stocked fishery. Allow the fly to sink and start a slow retrieve. As is usual, the fly came back and nothing of interest happened. We were disappointed, but why should we expect anything on a first cast? That would be too easy.

While Jeremy continued casting, I searched under stones for signs of life. I found a millipede under a stone and wondered why there are no fossil millipedes known in the Middle Old Red Sandstone of the Orcadian Basin when they are well known from the older Lower Old Red Sandstone of the Midland Valley. My reverie was interrupted by a wild expletive from Jeremy.

'Nigel! I had a touch—definitely a fish—Ah, it's on.'

'Well done! Try to keep it on.'

The instant appraisal was that this was quite a good fish as our fly rod bent in an attractive curve. After some nervous moments as it dashed about and splashed at the surface, a fish of about twenty-five centimetres was flapping at our feet.

'It's a *Cheirolepis*—very pretty.'

The fish was covered with tiny scales that gave an iridescent sheen on the flanks with dark green above and pale green to white on the underside. With great care and wearing gloves, the hook was extracted and the fish spooned to see what was in its stomach. Sure enough there were remains of 'shrimp-like' animals and conchostracans, but nothing else we could identify.

'Your turn,' said Jeremy, offering the rod, which I instantly put back in action. It was strange fly fishing at a time prior to the development of flying insects. There was no hope of a 'hatch' of flies, and we had not seen anything break the surface; there was nothing that might induce a fish to rise to the surface. I fished on in the normal sunk fly manner, trying different retrieves of the line to vary the depth and speed of the fly through the water. Jeremy kept up a chatter of encouragement interspersed with geological questions to which he usually supplied his own answer. The fishing banter also increased, with accusation of incompetence of ever catching the fish I had been studying for many years.

At last, a fish grabbed the fly. Although not large, a lively fish was on the line, darting backwards and forwards and trying to plunge to the lake floor. It was instantly recognisable—a small *Osteolepis* with thick shiny rhombohedral scales and a generally silver colour; it was strange to see a silver *Osteolepis*, as they are shiny and jet black when preserved as fossils.

We put the *Osteolepis* back, and the next cast produced an almost identical specimen, and then another. We had found a shoal apparently all of the same year class. After catching half a dozen we moved on round the shore towards the bay, watching the water and having an occasional cast.

'Change of fly time,' announced Jeremy, so we put a fishy lure on the end of the line and used a leader to resist toothy critters. The water was clear so a good plan was to drag the lure along about thirty centimetres below the surface like an injured fish, and hope something would attack from below.

This worked almost instantly: on the second cast, there was a big swirl in the water near the fly, and then another, but no take. Next cast the same thing happened; this was exciting but not productive. Casting out again I let the line sink whilst undoing a crossed line in the reel. Before I had started to retrieve the fly, the rod was nearly yanked out of my hand. The fish felt heavy at first but rapidly became sluggish, and then dammit—he came off.

'I've lost it. Bother. You try, and just let the lure sink,' I muttered. Over the next half hour we lost a further four fish and caught another *Osteolepis*. The *Osteolepis* felt quite different on the line to the mystery fish, and we were frustrated by our continual losses. Examination of the fly showed it was deteriorating rapidly; whatever we were catching (or not catching) was chewing up the fly something rotten.

'We could try a bigger hook, or even a smaller hook on the same size of fly. I think it is something with a very bony jaw.'

We used the smaller hook, trying to give the fish less leverage on the hook, and to seek out any fleshy areas. Another attempt. The losses continued, but at last we had one that stayed on the hook for a few minutes; it specialised in short bursts of speed, interspersed with dogged resistance. Gingerly we brought it into the shallows and immediately it was recognisable by its general ugliness. A large bony head, with large eyes, was followed by a leathery-looking body with pectoral fins, a single dorsal and a narrow, almost eel-like tail.

'That's *Coccosteus*. It has jaws with sharp shearing plates that have been messing up our flies.'

The fly was lodged in a fold of skin in the corner of the jaw. As we brought the fish ashore rather ignominiously by dragging it on to the gravel (we did not have a landing net), the hook came out of its precarious hold, and the *Coccosteus* lay gasping near the water's edge. Then it was sick, gave a strong flip of its tail and departed back into the lake.

'Dammit, it's gone again, but at least it has left us a sample.'

The fish had brought up a partially decomposed *Mesacanthus* and patches of *Dipterus* scales. Together with the remains of its meals were small pebbles. Examination of the regurgitated stomach contents showed other bits of bone and scale apart from the *Mesacanthus* and *Dipterus*. These bits were abraded, presumably by the small pebbles which were also in the gut and acted as gizzard stones to help break up the bony material, or maybe acted as ballast to keep the fish on an even keel. The last theory, however, did not look too good as the fish had happily swum off into the depths.

'I reckon the *Coccosteus* are like piranhas and just try to cut bits off other fish to disable them before swallowing them. That would explain the damage to the flies, and we probably kept losing the fish because they were hooked in the bony plates at the front of the jaw, and the shearing action of the tooth plates easily dislodged the hook. Anyway, no need to try and catch another one of those, they look evil beasts.'

'Not half as evil as its relative *Dunkleosteus* that lived in Devonian seas; that was six metres long and could almost have swallowed you whole,' retorted Jeremy.

As we continued around the lake shore the hills gave way to the margins of the sandy bay we had seen from the ridge. The water was shallower here and the bottom generally sandy with ripple marks parallel to the shore. Our feet sunk into the fine sand and beneath the sand was muddier sediment. The rod case now came into its secondary use, as a coring tube.

The tube of aluminium was split down each side and taped back together; operation was simply a case of pushing it into the sediment and then extracting it with a column of sediment inside. The tube went in quite easily, but sticky mud layers made extraction difficult. Pulling resulted in one of us being sucked into the mud. After considerable tugging and waggling of the core tube, the mud eventually let go and the tube emerged with a rich squelch. A smell of rotten eggs followed from the hole, indicating reducing conditions below the surface.

We opened the core tube and cut the core down the middle with a wire pull. Now we could see a section through the recent deposits of the bay. The core showed alternations of thin sand beds with dark muds. The sand beds were rippled, and from the base of one a sand-filled crack extended down thirty centimetres through the core. The sand-filled crack recorded a period when the lake level had lowered and the sand and mud flat had dried and cracked in the sun. Later the lake level rose and sand filled the mud cracks. Clearly there were frequent minor changes in lake level, since similar cracks were present beneath other sand beds.

'Well, that looks pretty typical of the marginal areas of the Caithness flagstones where there are a lot of fossil ripple marks and desiccation cracks.'

Clearly the core was full of bacteria, so we had to wash it out carefully and remove the sediment after making a record of the cored sequence. Anyway, we did not want a dirty rod case. Our attention then returned to the water, which here was shallow with a gentle ripple on the surface, so observation of underwater features was difficult and moving objects were hard to spot. Wandering along the water's edge we found little of interest on the sand, just the occasional bit of drifted plant and no sign of fish.

Towards the end of the bay the water was sheltered by the headland, and we could see into the water. The sand flat clearly ended about twenty metres off-shore, and the lake bed shelved away. There was movement to be seen, in the form of muddy clouds of sediment being stirred up by something feeding on the lake bottom. The clouds puffed up off the bottom, and dark shapes emerged and then created more sediment clouds. One was coming towards us into the shallows, and soon it was obvious that we had found Hugh Miller's 'winged fish'—the *Pterichthyodes* (Fig 3.2). It was easy to catch with the net because it had no concept of boys with fishing nets.

'What a strange creature,' remarked Jeremy.

The *Pterichthyodes* just lay there, unable to do any more than waggle its tail and aimlessly beat its pectoral flippers against thin air. Its box-like body armour kept it upright but stuck on the spot where we had laid it. From a keyhole-shaped opening in its head two pink piggy eyes gazed blankly upwards to the sky. Jeremy gently picked it up, and its flippers flapped pathetically in their ball-and-socket joints, and its tail poked rather rigidly out of its box-like body and wagged gently.

'Watch out for its dorsal fin spine. You never know, it may have been the stonefish of the Devonian,' I warned.

'Yes, I'm watching that, but it's lying flat. I think it's only there to support its dorsal fin.'

Just on cue the fish gave Jeremy a surprise: the fin spine erected to reveal a vivid yellow dorsal fin, and he nearly dropped the fish.

'Hmm, that looks like a warning if ever I saw one.'

Turning the fish over revealed that the underside was white, contrasting with the dark reticulate mottling of green and brown on the top. Its mouth was little more than a slit with lips that were gulping gently, and gill openings appeared as slits on either side behind the mouth. Several stiff rubbery feelers extended from the lips; they were waving about and were clearly sensory organs, probably used to detect food in the mud and sand.

'Jeremy, is its body armour rigid?'

'Not quite. The connecting tissue of the bony plates gives a bit of flexibility, but the plates themselves are hard but with a thin skin-like coating.'

Our fish was clearly not enjoying its unaccustomed trip out of water, so back in the net it went, to be returned gently to the lake. It recovered slowly and after a few moments swam ponderously away leaving an interesting track in the sand.

'That's an interesting potential trace fossil.'

'Yes, but I've never seen one in the rocks; in fact any fish trails are extremely rare in the lake deposits—it's a bit of a puzzle.'

As we watched our *Pterichthyodes* swim away Jeremy suddenly grabbed the net and swooped on a slight movement in the shallows.

'Gotcha,' he cried. 'It's a baby one—a baby *Pterichthyodes*.'

And so it was; hanging by a tiny flipper in the mesh of the net was a perfect juvenile *Pterichthyodes*, paler than the adult, and only a couple of centimetres long.

'It can't be very old, can it?' I mused.

'No, but it does show that *Pterichthyodes* must be breeding in the lake and growing up in the shallows.'

'Looks like they might grub about along the shoreline for organic material washed to the strandline—decaying plant bits and so on.'

The baby was duly returned, and careful observation showed that there were several juveniles in the shallows. They were virtually impossible to see unless they moved, for when at rest they shuffled into the sand and looked for all the world like a half-buried pebble. We soon realised that they stayed still if there was movement. All the eyes were programmed to do was to detect movement and instruct the fish to go to ground and lie doggo. The eyes also probably controlled the animal's daily rhythm by the simple detection of night and day. With eyes on the top surface, and mouth like a vacuum cleaner below, the beast certainly could not see what it was eating, relying on touch from its feelers, or smell to find morsels of food. A modern bottom-dwelling ray has the same organisation, finding prey by digging in the sand or by smell.

'Well, enough of this pond fishing with a net—let's get back to the big stuff and the rod. We are clearly unlikely to catch a *Pterichthyodes* on a fly.'

The one fish we both wanted to catch was a *Glyptolepis*, the top predator in the lake. At the geological time of our visit the largest *Glyptolepis* species was some sixty centimetres long and would have weighed about 2 kg. Later, their immediate descendants were much larger, maybe double the length and much more than double the weight. No complete fossil specimens of these monsters have yet been found, but we can judge the size from isolated jaw bones and other skull bones occasionally found in the Upper Flagstone Group in Caithness. As top predator they are unlikely to have been abundant in the lake at any one time, but we had already seen one, so maybe we had a chance.

Since *Glyptolepis* could certainly be classed as a toothy critter, the tough leader material was required and a 'fly' that maybe imitated a small fish. Logic and experience told us that our predator would cruise the shore hoping to catch an unwary young *Dipterus* or acanthodian, but would not be adverse to much larger prey if available. Our best chance of success was to use a large, general, silver, flashy, fish-fry imitation about three centimetres long. The hook was right at the head of the fly since that is usually the point of attack, and most prey would be swallowed head first.

We took turns with the rod, working our way along the shore to the next rocky headland. We were casting out to the edge of the shallows and employing various styles of retrieve—let it sink, slow steady retrieve, retrieve in bursts, strip it fast through the surface—all well tried and successful with trout and salmon under various conditions. From a rocky outcrop above the beach I could look down on the water and watch Jeremy cast and retrieve. He was working his way towards a rocky area of lake bed when a dark shape emerged from the rocks and swirled past his fly, returning to the rocks whence it came.

'Jeremy! Same cast again but three yards to the right and let it sink before you retrieve.'

Jeremy followed the instructions perfectly and the fly sank from view beyond the rocks from whence the fish had emerged. I did not see the fish this time, but Jeremy certainly felt it.

'It's on, what do you think it is?'

'I guess it will be *Glyptolepis*,' I responded. 'I saw it inspect your fly on the previous cast. It came out of the rocky ground.'

'It does not seem to be big or heavy enough for *Glyptolepis* but it's quite lively.'

In a couple of minutes Jeremy had the fish at the side. It was not a *Glyptolepis* but a very fine *Gyroptychius* of about forty centimetres—a big one. It was instantly recognisable from the diamond-shaped tail, and although it had a concentration of fins at the rear end—like *Glyptolepis* or a pike—it did not have the bulk of *Glyptolepis*. Clearly it hunted in a similar way and had ambushed the fly as it passed its rocky home territory. The teeth were numerous, and small but sharp, not dissimilar to those of a trout, and trout are effective fish predators, being able to swallow a fish of over one third their own length.

So, it was back to the *Glyptolepis* hunt, and my turn with the rod. We kept walking around the shore whilst I tried to cover likely spots with the fly. As we approached the next headland, the lake bottom dropped away and deep water was within casting range. Surely big fish cruised the edge of the drop-off from the shallow waters near the lake shore searching for any unwary fish. Casting as far as possible, I let the fly sink for ages on the slow sink line and then began the retrieve. There were many false alarms; sometimes the fly snagged water weed and algae and came back with green streamers, and on several occasions it truly snagged the top of a rock and had to be yanked free with some force. Luckily the rocks were covered with an algae and a lump of brown rubbery algal mat was left on the hook; at least we were not leaving artificial flies in the Devonian.

We were about to feel guilty that angling was overtaking science when the fly stopped again—another rock? But this was different; the rock moved. A big fish can just snap at a fly and stop, but when it feels the hook there is usually a sudden reaction—at both ends of the line. The time lapse from feeling a solid mass to frenetic activity was very short. After a couple of shakes of its head, the fish set off with great acceleration, the rod bent into a half-circle and the reel screamed as the fish tore off line in a hectic dive towards the deep water.

'Hell's teeth, this must be big.'

'Can you hold it on the reel?' queried Jeremy.

'Certainly not, I am not in control of this one.'

Then quite suddenly the fish slowed and began a head-shaking routine, much the same as salmon do. Now I could regain some line, but slowly; the fish did not want to leave the deep water, and every time I felt I had it to the edge of the deep water it made another run towards the depths. After five or six exchanges, it was clearly tiring and we had our first view: a great swirl and a large tail broke the surface. The fish was at least at the surface, and we could tell that on our light fly rod we had something of about 2 kg. The fish was still most reluctant to enter the shallows, and several times when we had it close to the shore it found new reserves of energy and managed a short burst of speed. Eventually it was done and rolled on its side. It could then be brought to the beach.

'OK Jeremy—grab it when you can.'

'Where? I don't think you can pick this up by the tail like a salmon. I wish we had a big landing net.'

'Go for the back of the head and get your fingers down to the gill covers for grip,' I suggested.

'What if it has spines on the gill cover like an Australian flat-head?' countered Jeremy.

'It doesn't; trust the fossil evidence, but mind the teeth. There's plenty of fossil evidence for those.'

The fish was unceremoniously grabbed and hustled on to the gravel of the shore. It lay there gasping, a full seventy centimetres long and deep-bodied with dark back and bronze-green stripes on the flanks which faded to a pale belly. Gingerly Jeremy went to remove the hook.

'Hey, you were lucky,' he commented. 'The hook has just fallen out. One more run and the fish could have been off.'

'Skill, Jeremy, skill.'

'No, you were just very lucky.'

'OK, I'll buy the drinks.'

This was *Glyptolepis*, the largest predator of the lake. Its head was large, and the jaws could open wide to reveal a cavernous red throat. Needle-sharp teeth lined the jaws and as they curved slightly backwards, any fish would have great difficulty escaping. It was just as well the hook had fallen out; we would have needed a gag if the hook had been inside the mouth. There were also large tusk-like teeth in upper and lower jaws, so it could give a good penetration bite and hold on. The general impression both in shape and colouring was reminiscent of a pike, and indeed it probably had a similar lifestyle as a lurking predator catching its prey by ambush amongst the rock and weed of the lake margin. The broad tail combined with the anal and dorsal fins would give the fish great acceleration over a short distance, and the rapid opening of the enormous mouth would have literally sucked prey into the jaws. Fossils have been found with the tail of one *Glyptolepis* protruding from the jaws of another. A case of 'eyes too big for stomach', and resulting in death of both parties.

After some shamefully tacky 'fisherman with fish' photos our catch was gently returned to the lake and held upright until it recovered. It must have suddenly realised it was free. We were both leaning over watching it recover when it leapt into life, slapping the water with its broad tail and providing us both with an impromptu shower. Then it was gone, cruising back to the depths to resume normal business.

'Bet it'll be more careful next time.'

'Yes, it will probably warn its mates about time-travellers with streamer flies.'

Our time was just about gone, so reluctantly the rod was packed away and the reel returned to its sample bag, and we marched swiftly back around the bay and started the climb up the rocky ridge back towards the Bus.

'Hey, we have not caught a *Dipterus*, and it is the most common fish at Achanarras,' Jeremy sighed.

'Well, it's too late now, maybe they are mainly vegetarian and grub about in the mud and weeds. Not the sort of fish to take a fly, I suppose.'

'Next time we'd better bring coarse fishing gear as well, and try and catch the large acanthodians like *Cheiracanthus* and *Diplacanthus*.'

'An interesting thought,' I replied. 'But somehow I think that this is a first and last visit, and we need to revert to use of a hammer and chisel in Achanarras Quarry.'

Excursion 4

Mass Death in Fife

TIME: Late Devonian.

LOCATION: Dura Den, Fife.

OBJECTIVES: To examine the deposits of rivers and sand dunes, and find the cause of a mass kill of fish.

THE MODERN EVIDENCE: The sandstones with many crowded fossil fish discovered at Dura Den in Fife in the nineteenth century.

'Accidental death' or 'death by misadventure' is the likely verdict in court for the animals entombed at many famous fossil sites. Dura Den in Fife is no exception. Fossil fish had been recorded from this area from 1830, and they had been figured by Louis Agassiz in his famous monograph on *Poissons Fossiles du Vieux Grés Rouge* published in 1844. Illustrious geologists of the day took the trouble to examine the fossiliferous sandstones at Dura Den. Charles Lyell, author of *Principles of Geology* visited the area in the company of John Anderson in 1842, and Hugh Miller, author of *The Old Red Sandstone*, also paid a call.

On 16 September 1858, however, the most sensational finds began, when Sir Roderick Murchison, John Anderson, Lord and Lady Kinnaird and a 'distinguished party from Rossie Priory' discovered a complete *Holoptychius* one metre long, together with many smaller specimens. Anderson, in his 1859 monograph on Dura Den, also recorded that the landowner subsequently cleared an area from which nearly a thousand fish were 'lifted from their stony bed of ages' (Figs 4.1, 4.2 and 4.3). Anderson dedicated his monograph to Lord Kinnaird and Sir Roderick Murchison, and Lady Kinnaird drew the originals of some of the plates, but they are more artistic impressions than anatomically correct drawings.

Twenty-four years before the finding of the *Holoptychius* bed, and fifteen metres higher in the rock section, a bedding surface had been identified bearing the remains of creatures of a distinctly problematical nature at the time of discovery. They looked like armoured boxes, covered by bony plates with a distinctive wrinkled ornament. The head, also encased in bony plates, had a single oval

Fig 4.1 Specimens of *Holoptychius* crowded together in a block of sandstone from Dura Den, Fife. Specimen in St Andrews University Collection. (Photograph courtesy of Stuart Allison.)

Fig 4.2 Reconstruction of *Holoptychius*.

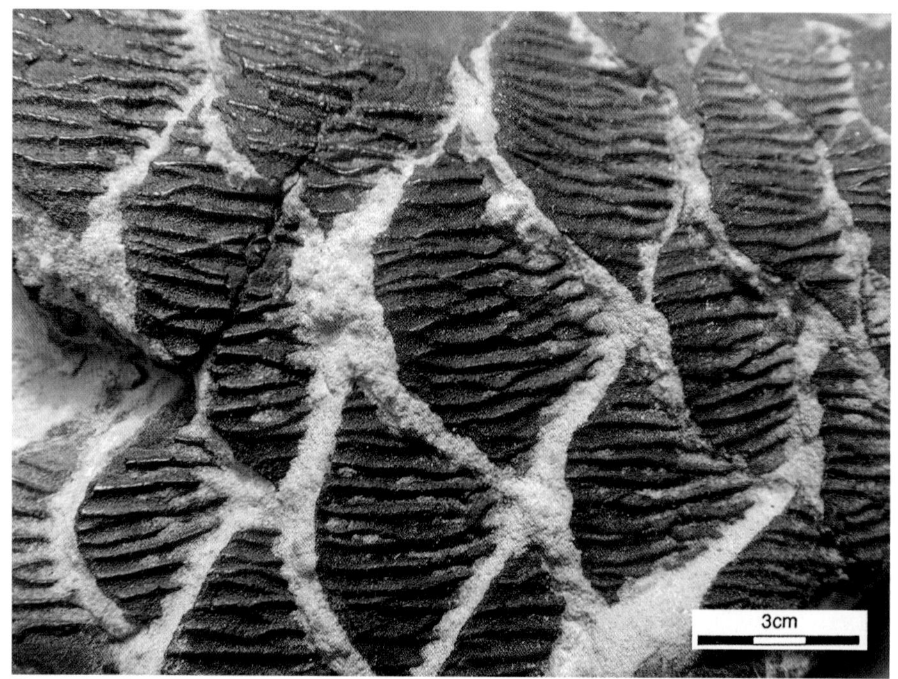

Fig 4.3 Detail of characteristic ornament of *Holoptychius* scales in a specimen from Dura Den. Individual scales are 30 mm in diameter.

opening like a cyclopean eye gazing upwards, but even more extraordinary were the two sickle-shaped arms attached either side at the front of the body. Rather poorly preserved was the part of the animal that gave the game away; a scaly tail with dorsal and caudal fins emerged from the rear of the armoured body, proving wrong those who considered it to be a crustacean. Underneath the head, weak jaws surrounded an opening that had to be the mouth.

Here was a fish the like of which does not exist today. *Bothriolepis* is now its name, but it was called *Pamphractus* or *Pterichthys* in 1859. In general body plan it clearly resembles *Pterichthyodes*, the 'winged fish' of Hugh Miller. The 1858 discovery was more spectacular than the *Bothriolepis* bed. Here the fossil fish were found crowded together on a sandstone bedding plane; the commonest fish was *Holoptychius*, a lobe-finned fish with large sculptured scales and strong jaws with pointed conical teeth—clearly of predatory habit. Along with the *Holoptychius* were other types of fish: a lung fish, *Phaneropleuron*, distantly related to the surviving lung fish species in South America, Africa and Australia. There was also *Eusthenopteron*, a famous fish from the zoological standpoint. Within the lobe fins of this fish there was a skeleton with bony elements arranged in a similar manner to the bones in the limbs of later tetrapods—four-legged animals that walked on land. Although arguments persist on the details, *Eusthenopteron* is clearly very close to the ancestral line of land-dwelling tetrapods. The most famous and best-preserved examples of *Eusthenopteron* came from lake deposits of similar age at Escuminac Bay in Canada, where they can be seen displayed in the excellent Miguasha Museum. Go there if possible.

The surviving relative of this group of fish, known as the crossopterygians, is the coelacanth *Latimeria*, the 'living fossil' discovered off South Africa in 1938. *Latimeria* is probably the last survivor of the coelacanths, a group of fish first recorded in the mid Devonian.

Slabs of the Dura Den fish beds are preserved in several museums including The Natural History Museum in London and The National Museum of Scotland. There are also two rather sad slabs at Aberdeen University, which were formerly mounted in the wall of the teaching lab in Marischal College either side of a small ichthyosaur from the Jurassic. The slabs suffered an act of vandalism some time prior to 1967 when the then professor (a mineralogist) ordered the laboratory attendant to clean the lab, and give the wall-mounted specimens a good scrub to remove the chalk dust and grime that inevitably accumulated in a teaching lab in the chalk-and-duster era. The attendant went to work with great zeal, but unfortunately used a wire brush in an attempt to remove the black bits and reveal the yellow sandstone. The black bits, of course, were the fish scales. Despite the damage the slabs clearly show that the fish are in part preserved in 3D, the round body shapes being partly filled with sand. To achieve this preservation the fish carcass must have dried out to a state hard enough to have allowed sand to enter the cavity within the carcass after the flesh had rotted away.

In the sequence of sandstones at Dura Den there are several levels with desiccation cracks—evidence that there had been water present, but that it had dried up through evaporation. Also within the section are sandstones of fine grain size with the grains well sorted. These are sands sorted by the wind as they were blown across a land surface. Thus we have evidence of fish dying in large numbers in an area where water was apparently only intermittently present.

The fossil fish of Dura Den occur within rocks of the Upper Old Red Sandstone of the Midland Valley of Scotland, and were deposited following a period of erosion that took place in the mid Devonian. Whilst Lake Orcadie was in existence in the north of Scotland, and the Caithness flagstones were being deposited, the Midland Valley was suffering lateral compression. Consequently the rocks of the Lower Old Red Sandstone were folded, faulted, uplifted and subjected to considerable erosion before the start of Upper Old Red Sandstone deposition.

The Upper Old Red sandstones were therefore deposited on an eroded surface of folded and faulted Lower Old Red and older strata. The unconformity between the two can be seen in the cliffs at Arbroath. Coarse-grained sandstones and conglomerates, representing the gravels of alluvial fans, accumulated where rivers dropped their loads of transported debris at valley margins. Broad alluvial plains built up in valleys as the old land surface with its topography of hills was drowned by the accumulating sediment. A large river flowed through the Midland Valley towards the North Sea, depositing its load in the form of large river sand bars. From the size of these sand bars we know that the river channel was at times about ten metres deep—clearly a substantial river system well supplied with water, possibly gathered from a distant mountain source. Within the drainage catchment of the river there was sufficient rainfall to supply the river and maintain a flow.

On the other hand there is evidence of dry, semi-arid conditions. Fossil soil profiles indicate that evaporation was greater than precipitation at the site of formation of the soil. These are caliche soils, with nodules of calcium carbonate and even laminated carbonate crusts. Such soils are found today in semi-arid regions of Australia, Africa and the Indus Valley. There really is no conflict here. The water in the river was derived from a distant source and flowed through an area with a local semi-arid climate.

Thus the Upper Old Red Sandstone deposits started with deposits of rivers and alluvial fans where water was brought to an area with a semi-arid climate. For a modern example of such a situation, we need to look at the Nile, sourced in the highlands of Africa and flowing through arid desert to the Mediterranean.

As time went by in the late Devonian the climate of Scotland progressively changed, becoming more equatorial and with increased rainfall. At the same time the sea was flooding Britain from an ocean to the south and extensive shallow shelf seas were being created where there had been land before. The great sandy outwash plains of Old Red Sandstone in south Wales and Ireland were inundated by the sea, which by early Carboniferous times reached the Midland Valley of Scotland.

This was not a gradual one-way flooding, like the filling of a bath, but a pulsed process with successive periods of marine advance working their way farther and farther from the open deep ocean to the south. The sea advanced and retreated many times, responding to an external control. Many consider that long-term climatic variations, climatic cycles, were led by the regular periodicity of the orbits of the Earth, Moon and Sun system. This climate control is known as Milankovitch cyclicity with periodicities of around 20,000, 40,000, 100,000 and

410,000 years. The reason for mentioning this again is to stress that climatic variations on Earth are dramatic and geologically very rapid. The most recent Ice Age ended only about 12,000 years ago. Note that 'most recent' is not necessarily the same as 'the last'.

Armed with this evidence, part fact and part interpretation, I was in a position to visit the Kingdom of Fife some time in the late Devonian just prior to the incursion of the Carboniferous seas from the south.

As the Bus tracked back in time there were excellent views of times of uplift and erosion of the Scottish Midland Valley. The Bus took me back to the early Devonian, and I watched the changes in the landscape below from a height where the whole of the eastern end of the Midland Valley was in view. Surprisingly the uplift period that resulted in the unconformity between Lower and Upper Old Red Sandstone could be tied down to only a few million years of rapid tectonic activity. The uplift was followed by a long period of erosion before subsidence started and the deposition of the Upper Old Red began. Faults that were active and clearly visible ten million years before were now inactive and blanketed with sediment.

My plan had been to accurately monitor the marine transgressions and regressions that took place in the early Carboniferous and trace shoreline positions at times of maximum inundation of the land by the sea. However, while the Bus took me through the late Devonian, which was still supposed by many to be a boring time of semi-arid climate, there were great variations in the amount of greenery and water in the general scene. This looked too interesting to ignore; there seemed to be clear climatic variations prior to the time the Carboniferous seas reached Scotland. Since the Upper Old Red Sandstone was usually considered to have been deposited in a semi-arid climate, it would be good to take the Bus down for a closer look at the land when there were flowing rivers and patches of greenery. However, there were still large areas of brown sand without obvious sign of vegetation. I chose a landing site on a flat area not far from a river channel (Fig 4.4).

———

Dust from the dry surface enveloped the Bus as it came to rest. As it gradually settled and drifted away on the breeze I found myself on a flat river plain with sparse low vegetation, much of which appeared to be well dried. The heat hit me as I stepped on to the surface—33°C was registering and visibility was poor due to dust and heat haze. It was the sort of scene in which a large red kangaroo would not have seemed out of place. The soil was sandy, but the surface hard and

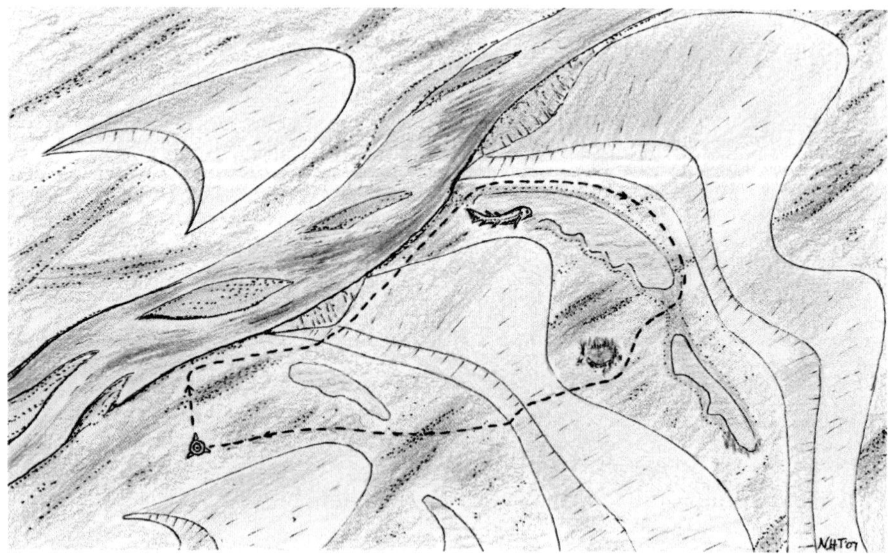

Fig 4.4 Route of Excursion 4 in a landscape of sand dunes, rivers and sandy outwash areas at Dura Den.

crusty; fragments of cemented sand were scattered on the surface and sheltered small drifts of sand blown by the wind. I was walking on a typical caliche type of soil, and conditions looked as arid as central Australia.

A further similarity with Australia was the sand dunes. They were crescent-shaped, generally less than ten metres high with their slip faces to the west. Thus the wind that formed them generally came from the east. The dunes merged in places to give long sinuous crest lines. The mobile dune sand was a pale yellow colour, contrasting with the more orange to redder colours of the soil surface.

I walked away from the Bus towards the river; this was clearly a major channel system and a complex of channels and sand bars stretched away into the distance. The channels individually were up to 100 metres wide, and there was a steady flow of water with a pale brown tinge. The river was flowing to the east. It was only possible to guess the depths of the channels, but judging from the sides of the exposed sand bars and the steep banks together with the steady swirl in the water, the water could well be more than two metres deep. It was certainly too dangerous to start wading, since the river bed was concealed by the murky water.

Wandering downstream I explored some of the sand bars that were attached to the main bank. It was obvious that the water level had been falling, since there were ripples on the bar surfaces and a series of terraces around the bars recording the falling water levels. Judging from the banks, the level had recently been almost a metre higher—this was a river subject to floods. There was no sign that the weather had been wet in the area I was inspecting, so this river fluctuation must reflect variable rains much higher upstream, in mountainous areas feeding this great river.

Continuing downstream, my path was interrupted by a sand dune that had migrated to the river's edge, the slip face falling direct into the river, which swept the sand back downstream. The sand was being washed down the river, deposited on sand bars and river banks, and then blown back up the sloping alluvial plains in the form of sand dunes. It was an easy matter to climb the dune face and peer over the crest. Here was a rather different scene: several shallow stagnant pools lay between this dune and the next. The pools were shrinking, for at the margins were patches of dried and curled mud, but there was something else that grabbed my attention.

It smelt awful. The stench came on a gentle but hot breeze and caused instant retching. Rotten fish, very rotten fish. The smell was the same in the Devonian as in modern times—quite unmistakable. Along the edge of the pool hundreds, if not thousands of carcasses lay in groups, often in shallow depressions, packed closely together, rotting and drying under a hot Devonian sun. Because of the stink, the fish carcasses had to be examined from upwind, so a brisk walk to the far end of the pool was essential.

From my new vantage point breathing was more pleasant, and the cause of this mass death became apparent. The pools lay in hollows between the sand dunes, and the water had come from the recent river flood which had broken through into the low area and flooded the hollows between the dunes. The fish had maybe sought shelter in this calmer, less sediment-laden water during a recent flood, but had failed to escape back to the river channel as the water level fell. The escape route had been blocked by wind-blown sand; maybe a gale from the east had blown sand into the narrow flood channel faster than the water flow could transport it away, and as the water fell the ponds became cut off from the river channels.

The fish were then doomed, and as the floodwaters fell further the water table lowered, the ponds dried out and the fish died. They lay in various states of decay, but generally with mouths open as if in a last gasp for air, and the bony scales supported the carcasses which dried in a semi-mummified form in the sand. In some cases the body cavities had burst open, the flesh was dried and shrunk, and the sand had trickled into every cavity of the dead fish and gently enclosed delicate dried fins. Indeed, along the margins of the pool wind-blown sand was gradually covering the whole of this scene of carnage.

In remaining puddles there was still some life, as fish gasped air at the surface; these were lung fish, hoping for a miracle of another flood, but still with a trick to try. Where the sand was soft and wet some fish were thrashing about in efforts to burrow into the sediment. When fully buried they would curl up, secrete a membrane around themselves, and enter a state of aestivation, a form of suspended animation, the body functions slowing to a virtual but not complete stop. In this state they could survive for up to a year, emerging when the pool was next flooded. There was surprisingly little evidence for any scavenging of this great mass of organic material; it seemed that few animals ventured out of the wet areas in this hot climate.

Such mass mortalities of fish are a common feature of arid areas. An example from Lake Eyre in central Australia was mentioned in Excursion 3: in that case fish were swept into the lake with floodwater from a distant source, and flourished briefly before being killed, possibly by rising salinity, as the lake dried out. Another modern fish mass mortality was seen by my Aberdeen University colleague Adrian Hartley in Namibia, where a coastal lagoon was lined with dead fish drying in the sun (Fig 4.5). There was no outlet, and the fish had been trapped. Blown sand was gradually covering the carcasses, as had happened at Dura Den. The only difference was that the Namibian fish had been partly scavenged by birds. Such events might be fairly common, but the chance of preservation in the fossil

Fig 4.5 Dead fish at the sandy margin of a coastal lagoon in Namibia. (Photograph courtesy of Adrian Hartley.)

record is very low, and such mass mortalities of fish preserved in sandstone are very rare indeed.

The surfaces of the sand dunes were bereft of signs of life; there was no evidence of burrowing animals and no tracks of arthropods or early terrestrial vertebrates. Yet there should be life, so where was it? It seemed that the shifting sands did not allow stable environments for terrestrial life to thrive and develop. Maybe seasonality or overall aridity of the climate was to blame, but there just did not seem to be a population of scavengers to clear up the mess. Later in time there would be amphibians, reptiles, birds and mammals to consume and disturb the carcasses, but here in the late Devonian they just remained, dried but entire, until covered by blowing sand.

I had seen enough and time was nearly up—the heat and aroma were incentives enough to return to the comfort of the Bus. As I trudged back, trying to avoid the soft dune sand where walking was more of an effort, I found some more permanently damp hollows in which there was some vegetation, and from the vegetation I flushed a few animals including a very cross and fast centipede. Since it had big fangs, it was a carnivore, so there had to be some prey about. All the animals were probably hiding in burrows and under stones, as happens in semi-arid areas today. There was also more dead vegetation than I had noted before, and searching and a bit of digging did reveal burrows in the surface over which the mobile dunes were migrating. I had arrived in the hot season; maybe the desert floor came alive with plants in cooler times. Maybe if I had returned after dark, with a torch, I would have found a variety of arthropods such as scorpions that emerged in cooler conditions, but for the moment I was the mad dog out in the heat of the noonday sun. I had had enough of the heat.

Once safely back on the Bus, I pondered on the likely scene when the fish of Dura Den died, and were subsequently buried and preserved. Chemical conditions in the sandstone allowed the preservation of the phosphate in their bony scales, and the sand grains of the Devonian dunes and the river had become cemented together to form sandstone during burial. Hundreds of metres of Upper Old Red Sandstone sediments accumulated during gradual subsidence of this part of Fife. A few million years later Carboniferous seas flooded the area and more sediment accumulated until the fish of Dura Den were buried two or three kilometres below the surface. It was at this depth that sand turned to sandstone.

After being buried for some 360 million years, and surviving periods of folding, faulting and uplift of the rocks, the Dura Den fish have come full circle, back to the surface, but now in an area of land where erosion is taking place. But it had to be man, in the form of a quarry worker, who, in pursuit of building stone, split a block of sandstone and revealed the marvels of the Dura Den fish to the Victorian world of scientific curiosity and enquiry.

Sadly there are no quarries working the sandstones today, and the chance of finding another great mass-mortality bed of Devonian fish in this area is consequently remote. Others probably exist, hidden beneath grass, soil and rock. Maybe the cutting of a new road or a deep pipeline trench will one day reveal a new Dura Den.

Excursion 5

The Trilobite's Tale

TIME: Early Carboniferous.

LOCATION: Bishop Hill, near Kinross.

OBJECTIVE: To visit the environment inhabited by some of the last of the trilobites living around a reef mound in a warm shallow sea.

MODERN EVIDENCE: The fossils of trilobites and other marine animals that are found in the tips at Clatteringwells Quarry on Bishop Hill, Kinross.

Trilobites must rank as the most popular small fossil. The reason that so many people are turned on to trilobites is that the whole animal is frequently preserved, and it clearly has a head, body and tail; but above all most trilobites have a head with two large eyes. The eyes give the different types of trilobites distinctive faces that excite the imagination. There is always more enthusiasm in a class for a discussion on the life habits of trilobites than there is for the (frequently) boring bivalves, or the depth to which a sea urchin might have burrowed in the sea floor.

The trilobites were a very successful group of animals, first found in the Lower Cambrian rocks laid down 545 million years ago, and dying out in the Permian about 300 million years later. Thus trilobites had a long innings on Earth; the mammals have a considerable time to go if they are to catch up. In their heyday trilobites occupied a great range of marine environments and developed many bizarre adaptations suitable to different lifestyles. Our interpretation of trilobite lifestyles is frequently only an educated guess, based on the shapes and habitats of modern arthropods with designs similar to trilobites. There are, however, some excellent studies on the details of trilobite form and function, notably the work of Euan Clarkson on trilobite eyes. He determined the field of view of a trilobite by detailed study of the direction of view of each of the lenses in the compound eyes of a group of trilobites typified by *Phacops*. This study showed that the beast in question had 360-degree vision in the horizontal plane and for about 15 degrees above its horizon (Fig 5.1). So this trilobite could sit on the sea floor and see all

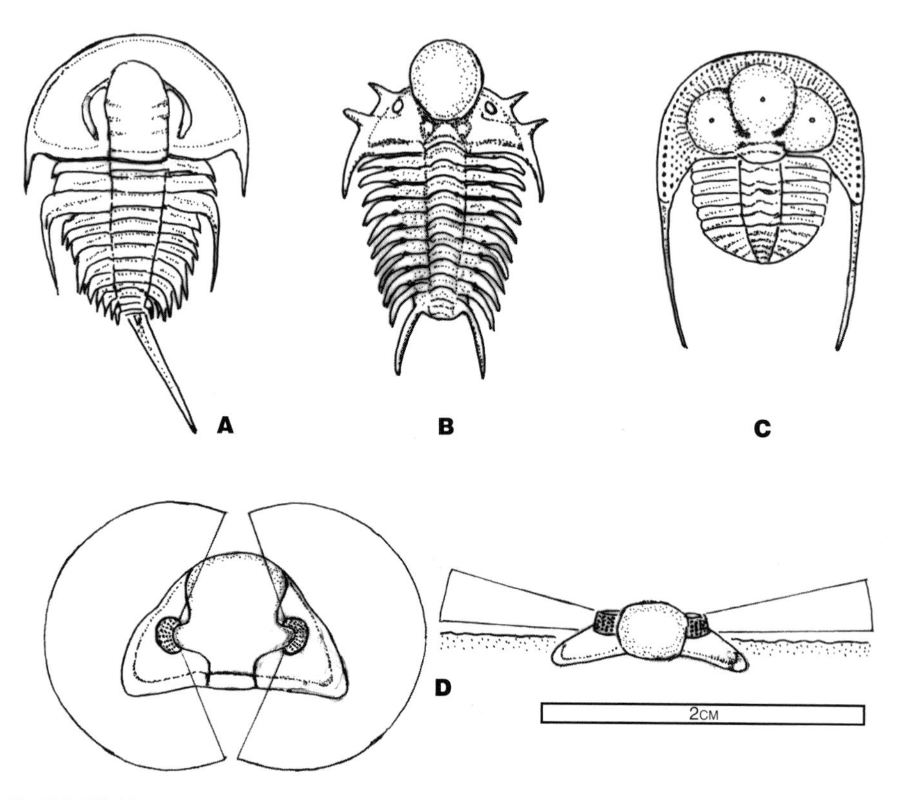

Fig 5.1 Trilobites. **A:** *Olenellus* from the Lower Cambrian of the northwest Highlands. **B:** *Sphaerocoryphe* from the Ordovician of the Girvan area. This spiny form with an inflated head may have floated in the sea. **C:** *Tretaspis* from the Ordovician of the Girvan area. This blind trilobite possibly burrowed in the surface of the muddy sea floor. **D:** The 360-degree horizontal field of view (left) and the restricted vertical field of view (right) of the trilobite *Phacops*.

that came within range. It could even lie in the mud or sand with its eyes poking out, and maybe set an ambush for passing prey. What it could not do was spot what was going on in the water immediately above. That is clearly a disadvantage if any of its predators could swim.

Amongst the variety of trilobite forms were some with a smooth, rounded body shape, but no eyes—they may have burrowed in loose sediment. However, other types without eyes had long hollow spines that would have prevented burrowing. These may have been filter feeders drifting in the oceans.

Spines on trilobites could have had more than one function. In types that swam freely or floated in the surface waters of seas and oceans, spines could have helped with buoyancy, whilst making the animal a rather prickly morsel for any hunter. Other spiny forms inhabited reef environments, and here spines and ornament acted as effective camouflage amongst corals, sponges and bryozoans.

When shape is combined with colour we can imagine that many forms would be superbly camouflaged. The wonderful shapes and colours of modern reef arthropods such as shrimps give some idea of the colours that trilobites might have displayed.

In Scotland, trilobites are found in rocks from the Lower Cambrian to the Carboniferous. The earliest, from the Lower Cambrian of the northwest Highlands, is *Olenellus* (Fig 5.1), a primitive type with many segments, still retaining features of the multi-segmented, worm-like organism from which all arthropods are thought to be derived. This fossil has an important place in the geological story of Scotland. *Olenellus* belongs to a distinct fauna that is also found in the Appalachians of eastern North America, providing evidence that the Appalachians and the Scottish Highlands are part of a single ancient mountain chain that was parted when the Atlantic Ocean opened only 65 million years ago.

Conversely the Cambrian faunas of the Highlands differ radically from those of England and Wales, providing evidence that these areas were once separated by an ocean. This was the Iapetus Ocean which gradually closed, welding parts of the UK together some 420 million years ago—the original 'Act of Union' for Scotland.

Trilobites are abundant in Ordovician and Silurian rocks of the Girvan area, with a great variety of forms, including some of the bizarre forms interpreted as having led a pelagic existence and others that lacked eyes and probably ploughed their way through soft sediment on the sea floor (Fig 5.1). Marine Devonian rocks are not found in Scotland, so there are no Scottish Devonian trilobites, but they do make a last appearance in the Carboniferous rocks of the Midland Valley. They are found in marine limestones and shales deposited at times of high sea level when the sea to the south flooded the tops of great delta plains that had been deposited by rivers draining the old Caledonian uplands to the north.

The subject of our story was one of the last of a long and proud lineage, an inhabitant of a shallow warm sea established in the Midland Valley during a period of high sea level in the early Carboniferous (Fig 5.2). The rocks in which this fossil is preserved can be seen on top of Bishop Hill in Kinross, and specimens may be found, with luck, in the spoil heaps of Clatteringwells Quarry, a long-abandoned limestone quarry.

Our trilobite is a mere twenty-three millimetres long, a far cry from the thirty-centimetre long and larger monsters of earlier times. It was known as *Phillipsia* for many years, so we can call it Mrs Phillips, although its name has now been changed by a palaeontologist to *Paladin mucronatus mucronatus*: so maybe it

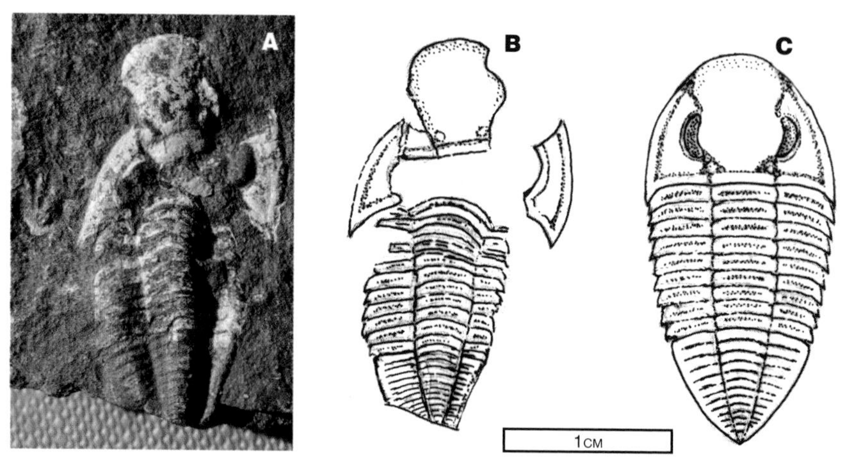

Fig 5.2 A: Fossil trilobite in Carboniferous shale from Clatteringwells Quarry, Bishop Hill, Kinross. The specimen is partly disarticulated, indicating that this might be a moulted specimen. **B:** Drawing of the specimen. **C:** A reconstruction of the trilobite; it was 23 mm long.

should be Mrs Paladin. However, I will stick with Mrs Phillips since *Phillipsia* is the name usually seen in the literature.

Mrs Phillips had a fine home, her residence for the last five years. It was one of the most secure sites in the area, lying deep in a crevice at the margin of a large mound. The mound was covered by a dense forest of seaweed that had strong holdfasts, anchored in the sediment. Mrs Phillips had previously had a rather more insecure home in a bare area of sea floor where there was little cover to hide from marauding fish and cephalopods such as *Orthoceras*—a straight-shelled relative of the modern nautilus. Now seven years old, she was a mature and experienced lady, having joined the annual spawning event for the past five years. She was feeling restless, the season was close, and on the next full moon under the cover of darkness the irresistible urge to wander and find a mate would overcome her normal caution. This was the most dangerous time for her, but yet it was essential for her to participate and continue to pass on her genes, and to continue the trilobite lineage.

She had found her present abode when its previous occupant—a larger female—had not returned from the spawning; as soon as Mrs Phillips had noticed the vacancy she took up residence. She had had to kick out a small male that had found it first, but he was not very strong and had been easily levered out and chased from the territory. This was prime territory, with lots of protective cover from seaweeds, crinoids and bryozoans (Figs 5.3 and 5.4). There was also plenty of food in the form of small creatures living in the sediment and hiding under shells of brachiopods (Fig 5.5), as well as the debris of crinoids (Fig 5.6)

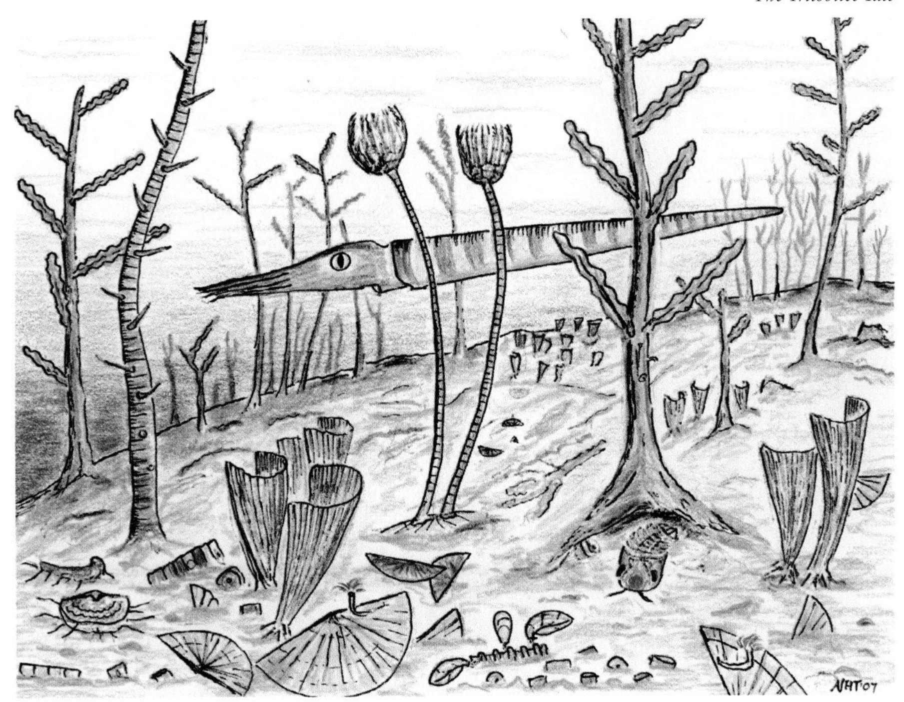

Fig 5.3 Sketch of the reef margin with a trilobite emerging from its hiding place. Bryozoans, brachiopods, crinoids and hypothetical weakly calcified algae clothe the sea floor. An orthocone cephalopod hunts for unwary prey.

that littered the sea floor. Every now and then Mrs Phillips had had to fight off potential invaders, but such events had become less frequent. There were not so many of her kind about.

Living in a secure nook provided a good measure of protection, and was one of the reasons why her kind of trilobite survived for so long, when many other types had disappeared. Life in the open was very dangerous for a small trilobite. Firstly there were the cephalopods: forms such as *Orthoceras* that hunted by sight, searching the bottom for any prey caught out in the open, or lying in ambush waiting for an unwary passer-by. However, these creatures had lived alongside the trilobites for virtually the whole of their existence and could hardly have been the main reason for their decline.

It was fish that posed such a problem for trilobites. They were now very numerous and diverse, and many groups had evolved and developed new feeding strategies in Devonian times. Fish were now much more efficient predators than the cephalopods; and indeed the cephalopods, like the trilobites, seemed to be in crisis. Rapid evolution of the bony fishes resulted in more efficient swimmers, faster and more manoeuvrable. Many were bottom feeders, sorting out small

Fig 5.4 Specimen of the fan-shaped bryozoan *Fenestella* showing the rooting structures that stabilised the colony on the sea floor. This Lower Carboniferous specimen was found at Bishop Hill, Kinross. Width of view 30 mm.

Fig 5.5 Rusty moulds of brachiopods that colonised the Carboniferous sea floor. This fossil from the Lower Carboniferous was found on Bishop Hill, Kinross. Shells up to 40 mm wide.

Fig 5.6 Crinoid stem fragments from the spoil tips of Clatteringwells Quarry. These are the most abundant fossils found at this locality, and represent several types of crinoid.

organisms from mouthfuls of sand plucked from the sea floor. The young of trilobites were highly vulnerable to this activity. Following a brief planktonic existence after hatching from an egg, the young trilobites colonised the sea floor, crawling amongst the grains of sand and the discarded and broken shells of brachiopods, molluscs and pieces of crinoid stems.

Another problem for trilobites were the sharks—not the sharp-toothed hunters that patrolled the open water picking off unwary fish and cephalopods, but those that had adapted to live on the bottom like the modern rays. They had jaws lined with flat crushing teeth, and found their prey by stirring up the sand and gravel, disturbing all manner of animals from their protective burrows and hiding places in the sand. Thus those trilobites that had relied on hiding in the sand surface and waiting for their own prey had now become the prey, and were hoovered up by the new forms of bottom-dwelling fish.

Now that Mrs Phillips was carrying a ripening egg mass under her body, she had no option but to find a mate in the annual spawning and have her genetic investment fertilised. Rather than chase away any male invading her territory, she now had to seek out a suitable mate. The only option was to leave the safety of her home and join the spawning party. The local spawning area was a patch of gravel made up of shelly debris; this provided a suitable substrate for digging a shallow nest in which the eggs could be laid, fertilised and covered to await hatching. It was the night of the spawning moon. Mrs Phillips cautiously emerged from her hiding place, the swollen egg mass making progress more difficult than usual. She part swam and part crawled over the bottom, instinct taking her to the spawning area.

There was not much time; she had to travel over a hundred metres, and at full speed it took fifteen seconds to travel a metre, so with diversions and rests it was going to take well over half an hour. It was nearly high water, so currents were slack. Luckily the sea was quite calm, so there was no wave action at the seabed. One year the spawning had been a disaster: a storm had blown up and many never made it through the currents, and those that did were picked off by fish or had their eggs washed away.

Mrs Phillips met a few other trilobites of her own kind on the way to the spawning ground. An old male, whom she had met a few years ago, sidled up hoping to renew an acquaintance, but he had been in a bad accident; part of his shell was bitten away, along with a number of legs. He could not keep up with her determined progress and fell behind in the darkness. Soon she had better company, a fit young male of about three years, smaller than herself, but in his prime. He latched on to the scent trail she was leaving and fell in behind her on the march. Soon another joined and competed with the youngster for the privilege of following directly behind her. A fight broke out, but Mrs Phillips kept steadfastly on her march. No point in waiting—the commotion could attract a predator, and anyway if the winner was worth his salt he would soon catch her up.

The young male trilobite was obviously a fit chap—before long he was at her tail again gently touching her with his antennae. He moved on to her back, clasping the top of her tail plate with his legs. They were now united, but she had to do more work to drag both of them along. Half an hour had now elapsed, and now she started testing the sea floor, digging little pits, and moving on, unsatisfied by the result. The scent in the water was now stronger; the mingled aroma of other couples that they bumped into in the gloom. Other males attempted to latch on to Mrs Phillips, but she had her man, and he flipped them off with his tail before they could gain a hold.

Now the digging became urgent. Mrs Phillips was happy with the sediment type, and dug herself into the sediment, dragging her mate down with her. When both were buried she released the egg mass, pulling her mate over the eggs, stimulating him to fertilise her production. They rested a while, buried in the sand, out of sight, allowing the few hundred eggs to be fertilised under the protection of their bodies. Her job was over for the year, she had justified her existence for the past year, but now the whole cycle had to start again.

Wearily Mrs Phillips and her mate emerged at the surface, he let her go and took his leave. He too had justified himself, but was not welcome company any more and so scuttled off over the sea floor. It was lighter now and there were other trilobites emerging and setting off for shelter, but they were not alone. Menacing dark shapes cut through the water above, intent on any movement below. Mrs Phillips sat still; her mate's progress was rudely terminated in a flurry of sand. The fish that had taken her mate turned and cruised towards her hiding place, hesitated and passed on.

It was an hour before she moved again, by which time the fish had gone, having captured many of her kind while others had fed on the eggs, sifting them from the sand. The tide was now on the ebb, and Mrs Phillips had currents to contend with on her journey home. She moved cautiously from one bit of cover to the next, only continuing when her eyes could not see movement, but her upward vision was poor and that was where the greatest danger lay. At least the water rocked the shells on the sea floor and created flurries of sand, so her progress was not as obvious as in conditions of complete calm. A small bottom-living shark appeared in her sight, small at first, but rapidly appearing larger and larger. She quickly dug into the sand and waited.

Crunch! The sand about her was forced into her side, and she was tumbling in the water. Instantly she curled into a ball, neatly tucking the rounded form of her tail into the purpose-made groove around her head shield. She hoped she looked like any other small pebble. The expected bite did not come; soon she was rocking back and forth on the sea floor in the currents of the sea. Cautiously she uncurled and checking that the coast was clear, continued on her way after her close encounter with death.

Eventually Mrs Phillips reached the relative shelter of growing weed and crinoids at the margin of her mound. She knew the geography now, and there were more places to hide. She was still cautious; some sneaky cephalopods with long tentacles lurked around the edge of the mound, and picked off passing fish and arthropods, including trilobites. Although in sight of her home, it was not the time to drop her guard. Whilst she had been away there could have been a takeover of such a good home. She was totally exhausted and hungry, but there was the chance that she would now have to fight for her abode. Cautiously she explored the entrance with her antennae; there was no sign of invasion, and no unfamiliar scent. In she scuttled, turning rapidly to face the entrance to her lair. All was quiet. A few morsels had drifted into her den while she was away and these provided her first meal after the spawning, the start of the annual process of building up strength and energy for nuptial events of the next year.

But would she be so lucky next time? Every year there were more dangers and fewer of her kind, and she was now an old lady. The odds were clearly stacked against Mrs Phillips, and against the continuation of trilobites.

Excursion 6

Volcanoes in the Jungle

TIME: Early Carboniferous.

LOCATION: Edinburgh and the Lothians.

OBJECTIVES: To examine the shores of the Carboniferous sea, and the swamps and forests beside rivers and lagoons, and to see the influence of active volcanoes on the forest.

THE MODERN EVIDENCE: The deposits of swamps, lagoons and forests, together with rocks representing marine invasions of the land area. Extensive floras and faunas such as those from East Kirkton. Volcanic rocks representing the roots of volcanoes as on Arthur's Seat in Edinburgh.

The dramatic skyline of Edinburgh, dominated by Arthur's Seat, Castle Rock and Calton Hill, is a product of the erosion of rocks of Carboniferous age. The hills are resistant igneous rocks, formed as the solidified contents of the vents of volcanoes, and the lavas that flowed from those vents. To these are added intrusive sheet-like bodies of igneous rock, with columnar cooling joints. The prominent cliffs of Salisbury Crags are an example of one such intrusive sill (Fig 6.1).

The softer rocks that have been eroded away to leave the spectacular hills were Carboniferous sedimentary rocks, mainly sandstones and shales. These rocks were ripped up by the passage of ice over the area during the glaciations of the ice ages that only ended some 12,000 years ago. Castle Rock is a classic 'crag and tail' feature with the 'tail', on which the Royal Mile is built, formed of soft sedimentary rocks, and the debris dumped in the lee of the crag from ice flowing from west to east.

Back in the Carboniferous, some 340 million years ago, the Edinburgh area was part of a large swampy jungle. Rivers meandered through the jungle and great lagoons formed near the coast which lay to the southeast. A large river delta fed by drainage from the northeast covered much of the area that is now Fife. Higher land—the Southern Uplands to the south and the Highlands to the north—was also clothed in dense vegetation. Scotland was close to the equator at the time and enjoyed a hot, humid, tropical climate, akin to that of the Congo basin in Africa or the rainforests of the Amazon in Brazil.

Fig 6.1 Oblique aerial view of Edinburgh looking east. Castle Rock is in the left foreground, and Calton Hill at the left edge, beyond the end of Princes Street. The eroded Carboniferous volcano of Arthur's Seat and the Salisbury Crags sill are in the distance. (Reproduced with the permission of the British Geological Survey. IPR/99-03CA © NERC. All rights reserved.)

The peaks of active volcanoes rose like giant pimples from the jungle, the tops of some rounded and vegetated while others were bare rock and ash with plumes of steam and gas rising from active craters. Steep ash slopes descended from the active peaks, and lava flows formed lobes on the lower flanks, leaving black scars where the forest had been destroyed and burnt by the lava. As a conservative estimate the volcano whose eroded and tilted remnants form Arthur's Seat was probably at least five kilometres in diameter and probably rose more than 900 metres above the level of the rivers in the jungle. Lava fields from the volcano may have linked with others in the area, and ash falls would have affected areas tens of kilometres from the volcano. Smaller and short-lived eruptions formed tuff rings where rising magma met shallow water and saturated sediment, resulting in explosive activity and volcanic ash formation. Such tuff rings would have contained lakes. If the activity continued and the cone built above water, magma would erupt at the surface to form lava flows.

Rivers flowing through the area became lost in creeks and lagoons before entering the sea through river distributaries. The whole area could be thought of as a large-scale river delta. The seashore was largely mud flats that merged with the vegetation of the delta; plumes of sediment muddied the sea where river distributaries entered it, and a few sand bars were formed at river mouths by the combination of river flow and the actions of tides and waves. Offshore to the southeast the sea was warm and shallow, and limestones and shales were deposited in clear water. The whole environment was dominated by rivers depositing sediment in the delta and at the shore, and there seemed to be little energy supplied by the sea in the way of wave and tidal currents. Hence the sediment was deposited near the river mouths rather than being reworked into broad sandy beaches, or swept away by the energy of the marine environment.

The Mississippi delta in the Gulf of Mexico is in a similar energy situation to the Carboniferous of the Midland Valley of Scotland. This delta has existed for millions of years, and the sediment represents the erosion products of a large proportion of the USA. Historically, the delta has built as a series of lobes several hundreds of kilometres across, each lobe containing river distributaries, and when these distributaries flood they deposit smaller lobes between the main distributaries, and so the delta builds. A new major lobe is initiated when the river finds a shorter and steeper route to the sea. Thus the river controls the delta, and the delta lobes only suffer minor reworking by the sea, resulting in narrow beaches and some offshore sand bars. The marine energy from waves and tides is very low in the Gulf of Mexico, resulting in minimal reworking of the delta front. As a contrast, the Senegal delta in West Africa suffers extreme wave energy and strong longshore currents. Here, the sand is reworked into a series of beach ridges more than thirty metres high, and the finer sediment is carried away to deep water. The Senegal delta receives as much marine energy from waves and tides in a day as the Mississippi gets in a year, including its short-lived but destructive hurricanes.

Deltas are low-lying areas, and hence are very sensitive to any relative changes in sea level. Sea-level variations can be global due to factors such as melting polar ice caps and fluctuating ocean volumes. Alterations in sea level can also be local and relative due to subsidence of the coastal area, rather than a global rise in sea level. Deltas are piles of unconsolidated sediment, and they compact under their own weight as water is driven out of mud and sand during burial. Futhermore, in large deltas such as the Mississippi the enormous weight of sediment concentrated in a relatively small area actually depresses the Earth's crust. Thus a continuous supply of sediment is required to fill the space left by subsidence, or else the sea will advance over the top of the delta. This is what is happening today in the Mississippi delta; land is being lost at an alarming rate, and man is the major culprit. Left to its own devices, the river would flood the delta and deposit sediment to maintain the delta, but man has interfered by controlling the river. Building artificial levees to stop the river flooding, and dredging the channel to enable big ships to sail into New Orleans, prevent sediment from being deposited on the delta top. Building dams for reservoirs on the tributaries of the Mississippi traps the sediment the river should carry to the delta. These actions are highly damaging to the delta and result in reduced flow and less sediment transport to the delta. The bulk of the sediment that does reach the delta is now lost because it is not deposited in the delta, but is transported within the main channels and disappears into the deep waters of the Gulf of Mexico. The great irony is that the USA is throwing its real estate into the sea.

Left to its own devices, the Mississippi would take a new and much shorter route to the sea down the Atchafalaya river. Indeed, some of the river flow is permitted (courtesy of the US Corps of Engineers) to take this route, and new land and a new delta lobe are being created where this river enters the sea. However, the Mississippi delta area is not a place to build permanent cities, as was tragically demonstrated by the destruction in New Orleans caused by hurricane Katrina in 2005. The area is ideal for a nomadic population moving in tune with the rhythm of the delta. In the eighteenth century the lesson was learned by the residents of the old delta town of Belize. It had to be abandoned due to flooding, and now lies under four metres or more of sediment: a dramatic illustration of the rate of subsidence and sedimentation in the delta.

My point is that in the modern Mississippi we can see the same natural forces at work that in Carboniferous times affected the Edinburgh area. There was a complex interplay between subsidence, sediment deposition and sea level. Sometimes sea level rose rapidly, the coast retreated and the sea flooded the land. At other times sedimentation kept pace with subsidence and a swampy coastal delta was maintained. When sediment supply exceeded that required to compensate for subsidence the delta built out, creating new land where previously there had been sea.

The present shoreline at Barns Ness near Dunbar in East Lothian, near the large cement works, is an excellent place to see the alternation between sediments deposited in the swampy delta and those of the marine invasions of the area. In the rocks exposed on the shore at low tide there are several limestone bands rich in marine fossils such as corals, brachiopods and crinoids. The sea in which they lived was warm and reasonably clear to allow coral growth. Periodically, deposits of the delta built out into this area and left shales and sandstones and even a thin coal seam with evidence for the growth of trees. So at Barns Ness we can see from the rocks that the sea advanced and retreated several times. Economically, this has provided good deposits for the cement industry.

At the present time there are serious concerns about rising sea levels in the world's oceans due to accelerating climate change. A rise in sea level of only a metre would swamp low-lying islands and cause rapid coastal retreat in low-lying coastal areas, particularly in highly populated delta areas such as the Ganges, and the aforementioned Mississippi delta. Millions of people who rely for their sustenance on the fertility of low-lying coastal land and the adjacent shallow seas could be displaced. The loss of coastal land as sea levels rise places great pressure on the population, leading to inevitable territorial disputes. Falling sea level may create more land, but results in loss and change of marine habitat and silting of harbours, and affects more severely those who rely on the sea.

The movement of the shoreline seawards or landwards is generally seen as a gradual affair, hardly perceptible in a human lifetime. However, there are many notable exceptions, even in the British Isles. In East Anglia rapid coastal erosion is taking place, with cliffs of soft sediment crumbling into the sea and rates of cliff erosion of more than one metre per year. The reason may be connected with climate change and rising sea level. Failure to maintain simple coastal defences such as wooden groynes, allowing protective shingle beaches to be swept away, and offshore dredging of gravel are more immediate causes. At Lyme Regis millions of pounds have been spent on cliff stabilisation, to the relief of property owners but not to the liking of fossil collectors who rely on fresh cliff falls and mud slides to reveal new treasures from the Lias. As a boy I can remember being impressed by the sight of half a tennis court at the top of the cliff between Lyme Regis and Charmouth; the net was still there, but it formed the fence at the cliff edge.

Erosion in southern England is nothing new. England is gently subsiding, and the coast has not naturally stabilised following the rise in sea level after the last glaciation. Ten thousand years is a blink of the geological eye. We should not assume that we can stop coastal erosion by simple coastal protection works, or that it would stop if the world reduced CO_2 emissions.

At some periods in Earth history it appears that sea-level change is rapid, possibly due to catastrophic collapse and melting of polar ice caps. The Carboniferous was a period during which there were rapid changes in sea level that can be traced in the rock record for thousands of kilometres, and possibly worldwide. Marine transgressions periodically invaded and destroyed the Carboniferous coal forests, converting forest to shallow sea. Detailed study of the fossils reveals that many of these events happened at the same time throughout western Europe. Thus this is a good period of geological time in which to investigate the natural effects of such sea-level changes.

One aim of our time-travel excursion was to study the Carboniferous coastal forests and swamps with reference to sea-level change. Our chosen objective was the Midland Valley of Scotland. Much of the technical part of the project involved recording detailed sequential images of the Midland Valley geography in the early Carboniferous to investigate rates of marine transgression and regression. From these images much information would be obtained regarding sediment input and deposition. Such data would be used in the better management of major delta areas of the world for the benefit of population and food production. Hopefully the lessons learnt by the disastrous treatment of the Mississippi delta could be avoided in the developing world.

We also needed some hard observational evidence from the ground, so we would visit and explore typical parts of the environment from coast to forest, and have the volcanoes and their interaction with the life of the jungle as a side issue. Our team was Hans, a palaeobotanist; Steve, who has a passion for arthropods; and myself for sedimentology. Three was deemed necessary because the jungle was probably dense and we might need to cut our way through the vegetation. So it came to be that we were all huddled over the monitor in the Bus with more than usual excitement and expectation as time scrolled back to the Carboniferous, and we stabilised position high over the future site of the capital of Scotland.

There were few safe landing places in deep jungle dotted with volcanic peaks, some high enough to be capped by clouds. The sea lay to the east, but there did not appear to be a solid beach; vegetation merged into water or exposed mud, and there was no pale stripe to indicate the presence of a firm sandy beach 'twixt sea and jungle. However, there were sandy islands a little way offshore, forming a broken arc. Waves broke on the seaward side of the islands, showing as a thin white fringe of surf. Landing options appeared to be limited to the offshore islands, and to sand bars that emerged from the waters of the large river that flowed through the area and seemed to lose itself in a maze of lagoons before reaching the sea as a series of smaller distributary channels. Then there were the volcanoes, some clearly active, with small plumes of gas and steam emerging from summit craters. There were areas that looked bare of vegetation on some volcanoes, but maybe these carried some risk in terms of stability for landing the Bus on rough surfaces, not to mention the outside chance of a volcanic experience.

It was clear that we had to choose the landing sites and the landing time very carefully to avoid disaster. Some views on the monitor showed jungle, wide rivers and lagoons, but no bare landing sites apart from the higher parts of active and recently dormant volcanoes. Any expedition would involve trekking downslope into the forest, probably hacking a path as we went, and with no guarantee that we would be able to force our way far enough to get beyond the area of direct influence of the volcano. It also appeared from the remote sensing that the forest canopy over wide areas was underlain by water. Maybe this was a flooded forest, and progress would be impossible without a boat. Wading and swimming were out. The inevitable risk assessment for the visit identified the very large predatory fish *Rhizodus* as a risk too far. We were all quite happy with this, and had no desire to splash about in murky water with something the size of a crocodile that had teeth like daggers.

However, the wet view of the area regularly alternated with a drier scene. This view from the Bus showed more bare surfaces: there were sand bars exposed in the rivers and along the coast. The lagoons had shrunk in size, and remote sensing did not indicate water under the forest canopy. There appeared to be less flow in the rivers, and the sea was only slightly coloured by suspended mud from the rivers. These transitions appeared to be rapid, on an annual basis, so it appeared that we were looking at seasonal flood events, probably caused by rains far away in the drainage basin of the river system. Possibly, the seasonal flooding that affects vast areas of the Amazon basin would be a modern analogue.

On a scale of thousands of years there were major changes to see in the geography. Most impressive were the advances and retreats of the shoreline. It was fascinating to watch the rise and advance of the sea as it flooded coastal lagoons, converted river mouths to wide estuaries, and invaded and killed the low-lying areas of forest. As the sea advanced, the dirty water of the river outflow decreased. Sediment was being dumped farther up the river system as the gradient of the river was reduced by the rising marine waters. Eventually the advance of the sea slowed and halted, and the river began to win the battle by depositing sediment

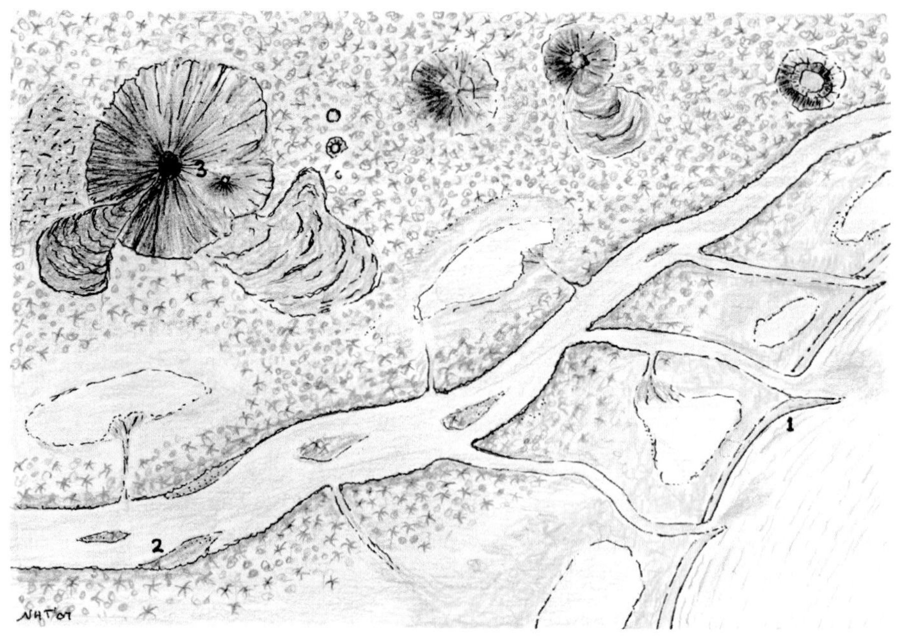

Fig 6.2 Sketch of the area for Excursion 6, with the three landing points.

to start filling the shallow coastal waters. Thus a delta started building seawards again, not as a smooth line of coast, but as a series of lobes. Each lobe was formed by a river distributary active for a century or so, before the river found a shorter route to the sea, and the old lobe was abandoned and a new one built.

We decided that the easiest time to visit would be when the delta was actively building into the sea, and during the season of low water flow in the river. With this combination of conditions, sand bars could be targeted as safe landing areas both in the jungle of the delta and on the coast. There was also a large volcano with a broad and reasonably flat area on one flank that looked a possible landing site. Remote sensing seemed to indicate that it was probably tuff rather than rough lava. There was a sharply defined crater near the top which showed as a bright hot spot on the heat monitor scan, but no signs of an eruption in progress. Of some concern was the clear presence of fire. Initially the smoke on the flanks of a volcano had been taken as evidence of active volcanicity, but the heat monitor clearly showed an advancing fire-front beneath the smoke.

Our first landing objective (point 1 on Fig 6.2) was a long spit of sand near the mouth of a distributary channel. This was the area that appeared to be most influenced by marine processes of waves and tides. From the air it was apparent that there was little flow entering the sea from this distributary, and the delta lobe was close to being abandoned. The lack of sediment input allowed the weak waves and tides to rework the delta deposits to form the sand spit and a narrow sandy shoreline.

The Bus landed with a firm jolt that reassured us that we were on firm ground. From the cramped confines of the Bus we stepped out on to the sand bar in bright sunshine. The air was very invigorating, and the view as attractive as any 'away from it all' location in a holiday brochure. Our sand bar curved gently seawards, lapped by the waves of a blue sea, and the only sound was the swish of the waves on the sand. Landwards the bar merged into a narrow beach fringed by dense vegetation. There were no large trees at the water's edge, but not far inland we could see giant lycopods looking like *Lepidodendron* and *Sigillaria* and horsetails (*Calamites*) the size of small trees at the edge of the jungle (Figs 6.3 and 6.4).

Wandering across the sand bar we rapidly reached the river channel. The water flowing to the sea was remarkably clear, and carrying very little sediment. Maybe this is the norm for the dry season, or else this is a dying distributary and the main outflow is elsewhere. We settled for a combination of both. Small terraces cut in the sand at the edge of the channel showed that the water level was falling.

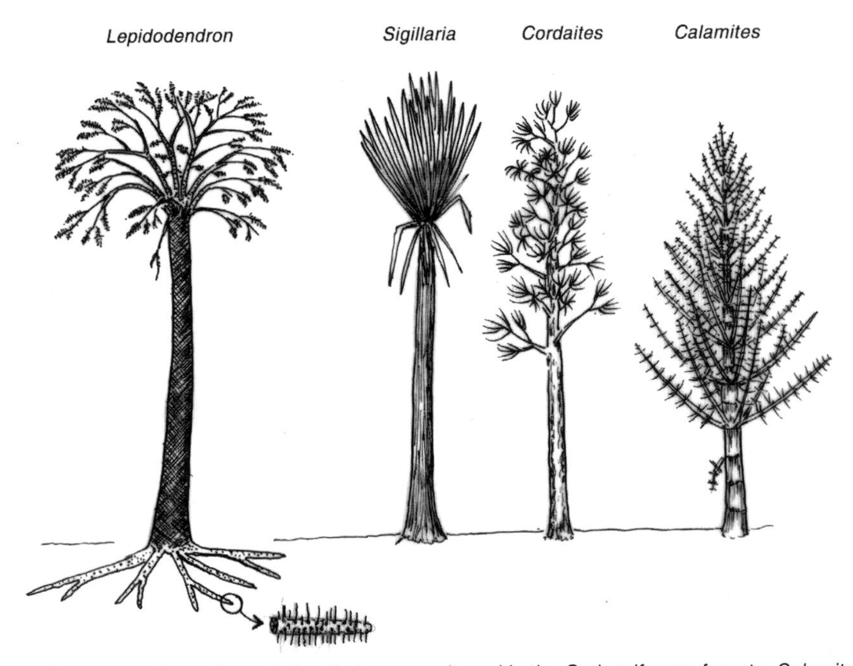

Fig 6.3 Reconstructions of vegetation that grew as 'trees' in the Carboniferous forests. *Calamites* was up to 18 m high, and the others grew to 30 m. An enlargement of *Stigmaria*—the rooting structure of *Lepidodendron*—is shown with the small roots growing from the main axis.

Fig 6.4 **A:** Part of a horsetail (*Calamites*). **B:** Impressions of *Lepidodendron* bark. **C:** Fossil root stem (*Stigmaria*) axis showing root scars.

Since there was no evidence of a flood from the river we took this to mean that there were tides, and that the tide was out. A quick check soon located the strandline from the last high tide, and we were pleased to note that the Bus was parked on top of the sand bar above the strandline left by the previous high tide. There is always something to find on a strandline so we wandered happily along announcing our discoveries.

'Horsetail stems, *Lepidodendron* branch and many seeds,' announced Hans.

'Baby king crab shell' was the first shout from Steve, and I picked up a rather worn coral, a solitary rugose coral not too unlike those fossilised at Barns Ness.

We soon had a collection lying on the beach that reflected the local environment and the transport of material to the shore. Pieces of wood, stems, leaves and seeds and even a tree stump reflected movement from the land. A few rather battered brachiopods, bivalves and some corals were part of the marine fauna that had been washed to the beach from offshore. There was also a living beach fauna; arthropods looking like woodlice or sea-slaters scurried away to hide when we lifted beach debris. They were using the strandline food resource in a similar way to modern sand shrimps that live in small burrows near high-tide mark. In the modern tropics burrowing ghost crabs are common on the upper shore, operating from deep burrows, but we are too early for them in the Carboniferous.

Then there was the sand to examine—a mix of quartz, feldspar and a little carbonate in the form of shell fragments. The coarser grains and the shell fragments were generally rounded by the wave action on the beach. When this sand is buried it will be porous and permeable and the shell fragments will almost certainly be dissolved rather than preserved as fossils. The general features we were looking at would not be out of place in Queensland, where the rainforest meets the sea and mangroves line the creeks. There the beach is frequently littered with leaves, tropical seeds, nuts and tree trunks, but there are also marine shells and fragments of coral from the offshore reefs.

Having reached the end of the spit we watched the waves and river currents mingling together. Such places are a favourite with fish that pick up morsels washed out of the river and disturbed by the surf. We could not see fish, but there just had to be fish there. We turned back towards the shore, and headed for the vegetation at the top of the beach. Hans was in the lead, eager to get his hands on a fresh green Carboniferous flora rather than look at fossil plants preserved as squashed, black carbonised fragments in grey shale. He was not disappointed and was soon regaling us with details of tree ferns, horsetails, lycopods and much more. He put names to many plants, and laughed and cursed at mistakes in the geological literature where a seed, leaf or shoot had been linked to the wrong plant.

'More names, more damned taxonomy to do,' Hans complained in good humour. Everywhere we looked there was research to be done. Only a fraction of the diversity of any ancient flora is preserved as fossils, so many plants were entirely new to science. It was also clear that the more robust ferns and woody plants could be identified, but of the smaller and more delicate plants most were new to Hans. They would be difficult to preserve in sediment, and so must be rare as fossils. The area behind the shore was low-lying and the vegetation just above head height; it had either not yet had time to develop into tall forest trees, or the environment was not suitable for large trees.

The onshore breeze rustled the ferns, the noise combining with the gentle lapping of waves on the shore. Conditions were perfect but very quiet compared to the present day. We were not aware of any animal noises, but surely the amphibians of the forest must have developed some means of communication. Modern forests are full of animals that communicate by sound, since visibility is so poor. Modern amphibians such as frogs and toads are particularly vocal in attracting a mate, so why not the same strategy in the Carboniferous? We needed to look closer, and since we were faced with a green wall of vegetation up to head height, there was no option but to cut our way in and make a track.

'OK Steve, let's break out the machete. I don't think we will alter the path of plant evolution by doing a bit of gardening.'

'Yes, but do give me some nice cut sections of the stems,' responded Hans.

Physical progress in the jungle would clearly be impossible without doing some damage. Hence Steve attacked the ferns and horsetails with enthusiasm as we made progress inland. The plants were growing on sand at the back of the beach, and our path took us downhill.

'You beauty.' Steve dropped the machete and made a dive to the vegetation below. He turned with his gloved hands cupped around some animal.

'Guess what I've caught?' Hans and I craned forward to look, and Steve opened his hands to reveal a spider with very long legs and a small body.

'Ah! A harvestman, and a big one, that is excellent,' enthused Hans. The beast crawled over Steve's hand and was gently returned to the vegetation. It looked like a giant version of a modern harvestman the reader might find in their garden. Harvestmen spiders have been around for a long time—Hans with Hagen Hass having found the oldest known harvestman in a block of Rhynie chert at the lab in Münster. These spiders have been stalking the undergrowth for more than 400 million years and have changed very little, despite big developments in vegetation.

'It looks clearer ahead,' observed Steve, but as he strode forward he suddenly appeared considerably shorter to his followers, and there was a lot of splashing while expletives filled the air.

'You OK, Steve?'

'Yuck,' came the reply, as major squelching noises preceded his emergence.

He had gone into swamp up to his knees, and as he emerged the smell of rotting vegetation filled the air.

'Reducing conditions then; effective demonstration, Steve.'

'I just fell through the surface,' he explained. 'There is a mat of vegetation with water and rotting vegetation underneath. I guess it is a floating bog that has filled the backshore lagoon.'

This effectively explained the lack of tall trees; they would have no support. We had worked our way down the back of the sandy beach ridge and fallen in the bog. As we could go no further inland here, we worked our way along the edge of the bog and soon found small areas of open water.

The first dip with our small pond-dipping net brought instant excitement. The net was half full of vegetation bits from the bottom of the pool, but there were a lot of creepy-crawlies to see. Most obvious were shrimps (Fig 6.5) which jumped in the net, as would modern shrimps from a seaside rock pool.

'Where is Euan Clarkson when we need him? Such an enthusiastic supporter of Carboniferous shrimp research would love these.'

'They look close to *Tealliocaris* to me, as in the shrimp bed at Cheese Bay, but are rather better preserved.'

'Are there enough for a prawn cocktail?' queried Hans, who always liked good food.

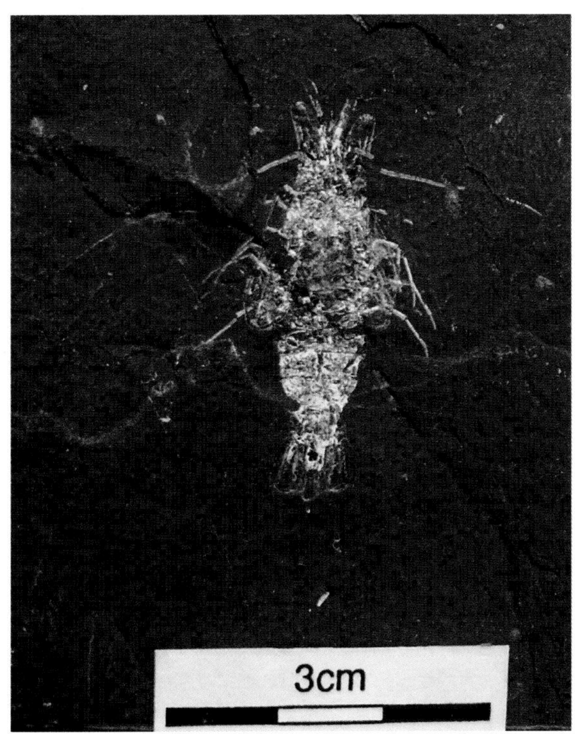

Fig 6.5 Carboniferous shrimp *Tealliocaris* from the Cheese Bay shrimp bed, West Lothian.

There was a lot more of interest in the net, including several larval stages of arthropods, and various worms. Steve tipped the contents on to the ground and started to pick off the leaves. Something was struggling to push off the wet vegetation. Out came an animal we could recognise—a dragonfly nymph. It was about seven centimetres long with an ugly face and a vicious extendable claw tucked under its head for capturing live prey.

'Oh! Here is its dinner,' said Steve, sorting a large fat grey tadpole with black eyes and feathery gills from the contents of the net.

'Must be the larval form of one of the amphibians. I hope we meet the parents.'

'Amazing. Dragonfly nymphs eat the tadpoles in my pond at home today, a relationship that goes back to the Carboniferous; little has changed in 350 million years for some animals.'

Having realised that we had spent far too long playing in the water, we returned the catch to the water. As we left the pondside Hans turned over a small log with his foot, and instinctively retreated as a large black-and-orange centipede about twenty centimetres long rushed out and headed for cover. Closer examination revealed a lot more life lurking under the log. There were other centipedes, millipedes and many worms crawling in the wet and rotting vegetation. Fungi were growing from the log and white meshworks of fungal hyphae invading and breaking down the dead plants. It was a familiar scene—the general structure of communities that live in decaying plant litter and the top few centimetres of soil has changed little since the Devonian, but here the biodiversity was far greater than we had seen at Rhynie (Excursion 2).

It was time to go back to the Bus; we hoped to visit two more places and we were already behind time. While returning the way we had come, we saw no animals but did hear a scurrying sound followed by a splash and silence. Nothing could be seen through the vegetation, but the noise told us that if it was an amphibian it was a lot bigger that a modern newt. Then we were buzzed from above. A distant hum combined with a brittle rattling noise became louder and louder. We ducked in unison.

'What was that?' I queried, having felt the draught as it went by.

Steve was craning his neck to peer over the vegetation as the noise faded.

'Dragonfly, and a pretty one: electric blue, with big black-and-white eye spots on its wings. Only about thirty centimetre wing span.'

'Only! That seems big enough,' retorted Hans.

'Yes, but some were nearer a metre wingtip to wingtip—the biggest dragonflies ever.'

'Why so big?' queried Hans.

'Oxygen,' said Steve. 'Carboniferous air had about 30% oxygen compared with about 21% today. That meant it was easier for arthropods to absorb oxygen through organs such as book-lungs and trachea, so they could grow bigger.'

'Makes you wonder what they eat, seeing as how modern dragonflies take prey on the wing. I wonder what else around here can fly?'

There was no instant answer to my question, as we had to make tracks to the Bus. Stomping back along the beach ridge we made good time, and had loads of energy for minor diversions, including washing the stinking mud off Steve's legs. That would make the Bus smell bad. We all felt very fit, almost high on excitement. This was probably also due to the high oxygen content of the air. We were on an oxygen trip.

For our next stop we were heading some 50 kilometres upstream to a sand bar at the edge of a major river channel (Fig 6.2 point 2). From the Bus it looked as though we would be able to get into the jungle beneath the canopy without having to swim or paddle. The Bus made a perfect landing. Outside it was hot and humid; earlier we had been cooled by the sea breeze on the coast, but this was not so pleasant. The sand was firm enough and covered in ripple marks left from the last time the river topped the sand bar. It looked as though that event was some time ago because there was a good scattering of leafy debris that had fallen from the trees, together with many signs of disturbance by animals. There were so many trackways crossing each other that it was difficult to pick out individuals. Imprints of feet were generally small, up to a few centimetres, but one was much larger. The impressions were the size of a large human hand, and between two rows of prints was a drag mark where the belly of the animal had touched the sand. Superimposed was a tail drag. The whole trackway was nearly a metre wide, and had to have been made by a four-footed beast about three or four metres long. Clearly there was a large amphibian for us to find, and maybe treat with respect.

We walked along the sand at the river bank. Overhead towered huge *Lepidodendron* trees, rising above horsetails (*Calamites*) and ferns that lined the river bank. The *Calamites* here frequently grew direct from the water, forming a giant ancient equivalent of reeds at a river bank. (Modern diminutive horsetails also grow in water, but are easily overlooked.) The river was wide, the far bank maybe a kilometre away. The stream flowed strongly but rather slowly. Underwater at the edge of the flow the sand was in motion and forming ripples that gradually moved downstream as we watched. Out in the channel the water swirled around low sand banks, and a few showed green where plants had gained a hold. Log jams of vegetation had piled up, giving some plants a precarious temporary hold; they were potential victims of the next seasonal flood. Trees had also toppled from the banks, and one now barred our way upstream. The crown of this *Lepidodendron* lay in the water, and sand had been scoured around it by the current. The trunk, some seventy centimetres in diameter, bore the artistic repeated diamond shapes of leaf scars; something normally only seen as fragments in fossils, but here covering a whole trunk. The tree had brought down many climbers that entwined its trunk; some of these were now dead, but one was still living and thriving in its horizontal position on the unaccustomed availability of light.

The trunk formed a convenient bridge to the bank, where its roots, the familiar *Stigmaria*, were upended in a fan (Fig 6.6). I put my hand on the log and bent forward to hoist myself up. Two large eyes met mine, but in a flash the eyes parted and a dragonfly clattered noisily into the air. It is bad enough when a small insect makes a surprise move, but worse for the heart rate when it is the size of a blackbird. The dragonfly flew off hawking along the river bank, coming to rest a few hundred metres away.

'It was just resting,' said Steve.

'Lovely plumage,' mused Hans.

'But certainly not an ex-dragonfly.'

The Pythonesque exchange relieved the tension.

'Maybe they feed at a particular time of day, and they are waiting for a hatch of flies.'

'Not more fishing stories,' pleaded Hans, 'and anyway it occurs to me that we are probably about 30 million years before the time of the oldest-known dragonfly.'

'Ha! Never trust the geological record. A few years ago the oldest harvestman spider came from the early Carboniferous, then you found one in the early Devonian Rhynie chert—about 60 million years older. So stretching the history of dragonflies back a mere 30 million does not bother me, especially as they have such low preservation potential.'

I walked along the log and jumped off on to the river bank. At first we peered into darkness under the canopy of the forest, but rapidly our eyes became accustomed to the gloom, and we found ourselves in a very different world. This was clearly mature vegetation. Tall straight trunks rose like cathedral columns. At the tops the crowns spread into a fan-vaulted ceiling, each fan merging with one from an adjacent tree. Light filtered down, but little grew on the forest floor. There was

Fig 6.6 The 'Fossil Grove' in Victoria Park, Glasgow, photographed in 1937. The stumps (*c*.0.4 m high) of large club-mosses (lycopsids) and their roots are preserved in a fossil soil. The exposure is now preserved inside a building erected over it. (Reproduced with the permission of the British Geological Survey IPR/99-03CA © NERC All rights reserved,)

a thick spongy layer of decayed plant litter. The dampness and humidity ensured that decomposition was rapid; no dry leaves to kick about. All was soft, with a distinct smell of fungi.

Digging down revealed a bit of a surprise: layers of vegetation and layers of sand. The only source for the sand was the river, so we had evidence that the river floods periodically spread sand into the forest. Therefore at some times of the year there was water swirling around the trunks of the trees. It seemed that the sensor on the Bus had not lied; water did flow through the forest in annual floods.

As we moved away from the river the going became wetter. Very gently we were going downhill and were on a vegetated levee. Digging here revealed less sand, and poking showed that the ground was soft, probably all accumulating rotting plant debris. Now we were seeing more dead and fallen trees. Trunks lay prostrate on the forest floor in all stages of decay, some so rotten that only a raised ridge with many fungi remained, and a few pieces of bark. When the trees fell they had brought others down, or snapped them off to leave broken stumps. A falling tree had left a gap in the canopy that adjacent trees had tried to fill. When sunshine reached the forest floor, new growth had started, long thin stems straining rapidly upwards towards the light, trying to be the first to claim the life-giving hole in the canopy.

Stopping at the edge of a small clearing to rest we perched on a recently fallen tree, its roots partly ripped from the forest floor. The sun streamed down sending shafts of light through the new growth in the clearing. Several large old tree stumps gave the impression that several forest giants had died to create the clearing, so triggering new growth.

'What's that? Something moved by that stump.'

'Which stump?'

'The one with the fern growing from the rim.'

'Let's go and look.'

'No, let's wait and watch for a few minutes,' I suggested. 'We have seen no animals yet by walking about.'

More movement, and cautiously a head emerged from the stump and appeared to look carefully about, before venturing forth. The head, broad and rather flat, was followed by a body similar to that of a newt, with the same rather spindly legs that did not look strong enough to support its weight. It climbed down the stump with remarkable agility and set off purposefully with a waddling gait, legs spread out from the body. This gave the body an eel-like sinusoidal motion as it pushed through plants and litter. Watching its progress was taking time, and we needed to explore more. As we got to our feet the amphibian sensed our presence and sped up, then it disappeared into vegetation and there was a splash. We followed and found ourselves on the bank of a muddy creek. It was almost hidden by overhanging and fallen vegetation, and we were certainly not going for a paddle. Our amphibian had vanished, only a few dying ripples remaining on the water.

'What would you call it?' asked Steve.

'Well, it's an amphibian, possibly a temnospondyl. *Balanerpeton* was the most common amphibian in the East Kirkton fauna from Bathgate, and grew to at least fifty centimetres. It could be that, but of course we would have to dissect it and look at the bones to make any comparison with fossil material.'

'Hmm, that mud looks well trampled,' observed Steve.

'Maybe there are lots of the amphibians in there, all being very still and silent.'

However, we were looking at the wrong scale—the mud in front of us moved just a bit. We all saw it and instinctively stepped back.

'That's a tail.'

'Well done, Hans. The head is at the other end. I'm glad it's pointing away from us.'

'It's practically invisible in the mud, but I can see an eye, about five feet from its tail.'

We watched for a minute but the creature showed no sign of action; time to help nature. A small branch dropped in front of it provoked no reaction. A small branch dropped on it; still no reaction.

'Nothing else for it,' I suggested. 'We need the longest stick we can find to give it a poke, Steve.'

'You can poke it yourself, Nigel,' he retorted. 'Try and remember your backwards long-jump skills.'

So that was decided. I poked it just behind the point where its front legs disappeared into the mud. The reaction was dramatic. A broad head with gaping red jaw thrashed round, and teeth crunched into the stick. At the same time the tail thrashed a shower of mud, spattering my colleagues. The beast let go and hissed, shaking the stick from its mouth and resuming its ambush.

'Well, that was exciting—a large predatory amphibian. I guess he's waiting at the water's edge for any fish or amphibian to chance by. Looks like the crocodile of his time. I have no idea what he, or she, is called. Too much mud to see any detail, even the natural colour is impossible to guess.'

We left our muddy lurker, and Steve and I started back towards the Bus while Hans had a look in the stump from which the small amphibian had emerged. He called us back.

'Look in here, it's laid eggs.'

'Really?'

Steve and I suspected a joke, but sure enough there in a small pool of water at the base of the stump were several typical amphibian eggs: pale jelly with a dark centre and each about two centimetres in diameter.

'She needs to spread her eggs to give them a better chance; maybe they would be eaten in the creek by fish or other amphibians. She is also breeding in the drier season. This place is all water in the wet season, and eggs would be washed away in the flood. That little pool would offer good protection.'

'I bet a few small animals fall in there and provide the tadpoles with food.'

'Maybe the amphibian fossils found in tree stumps at Joggins in Canada were doing the same thing,' mused Steve as we continued on our way.

We retraced our steps to the river and the tree bridge, and we were soon on the sand by the river. We were actually early, so it was time to reveal the length of monofilament fishing line with hook wrapped around a cork. There was only silver paper on the hook, but it might resemble a small fish. With the free end tied securely to a stick I sent it into the current and it swung round towards the bank. Nothing. I hand-hauled it back to try again, and fancied there was a small swirl in the water just as the hook left the water. Second try, and bang, something was on. It was only a tiddler, but still a nice shiny trout-shaped fish with the body scales extending into the upper lobe of the tail as in typical palaeoniscid fish. It had small sharp teeth, suited to many types of prey from worms to insect larvae and small fish. Maybe it was something like a *Cosmoptychius*, but since the fossils are only represented by fragmentary carcasses there was no way a positive identification could be made. Therefore the fish went back into the water. I was rolling up the line around the cork when there was a small splash on the water, and then another and another.

These were flies—or more closely the type known as mayflies, but much larger. A hatch was in progress and as the insects emerged through the water surface they were being snapped up by fish. Many made it into the air to mate, but there was further danger. Right on cue the dragonfly was back plucking, the large juicy mayflies out of the air and eating them on the wing. This was the lunchtime hatch of flies, and rather more exciting than the same event on the Aberdeenshire Don. I must remember to bring a fly rod next time.

Returning along the riverside sand bank to the Bus we saw more amphibians, but also had glimpses of a faster animal that looked like a small lizard. It was hunting insects, and would stand still and wait for one to come into range before snapping it up with an agile leap. It too was taking advantage of the mayfly hatch. This seemed to be similar to reconstructions of the famous 'Lizzie' fossil from East Kirkton. Hailed originally as the 'oldest reptile', it is now considered a proto-reptile, and is known as *Westlothiana lizziae*. The original specimen was found by Stan Wood, a well-known collector and dealer in fossils, and now (following negotiations and scares that it would be sold overseas) resides as a treasure of The National Museum of Scotland.

Near the Bus, Steve surprised an amphibian that was resting at the water's edge. It splashed clumsily into the water and swam into the river. It did not look relaxed and swam as fast as it could to regain the bank farther downstream. Suddenly there was a huge swirl in the water and the amphibian thrashed furiously towards the bank. It had nearly made it, but the water suddenly erupted and a gigantic head emerged with gaping toothy jaws and the amphibian had gone (Fig 6.7).

Fig 6.7 Reconstruction of *Rhizodus* attacking an amphibian in an early Carboniferous river scene.

'Wow! What was that monster? Did you see the teeth?'

'The head was enormous, and the mouth had very wide gape. That poor amphibian was only a snack; the creature must be able to swallow large prey.'

'Maybe as big as you, Hans,' joked Steve.

'I know what that was,' I interjected. 'Just the size and a view of those tusk-like teeth with smaller sharp dagger teeth between make it *Rhizodus*, probably the top predator of the rivers and lakes.'

'I guess it can grip a large amphibian with those tusks, drown it and then rip pieces off if it can't swallow the prey whole,' mused Hans.

'The lower jaw of a big one can be a metre long, and with the tusks more than twenty centimetres long,' I explained. 'I do not think I would fancy being in a canoe on this water, and risk being mistaken for a swimming amphibian. A modern analogy would be a great white shark mistaking a surfboard and surfer for a seal. Forget what I said about the fly fishing. I would not go in the water here in waders.'

We entered the Bus and set co-ordinates for our third locality of the excursion; we were off to inspect the slopes of a volcano. We arrived in what seemed a different world (point 3 on Fig 6.2). The Bus was perched on a gentle slope of reddish crunchy granules—clearly volcanic tuff. We were on the flank of a volcano, on a relatively flat area, but behind us rose the volcanic slopes, with only sparse vegetation. There was a lot of loose tuff, and some large blocks. Ravines showed where the soft volcanic tuff had been eroded by water flowing down the side of the volcano. The flat area chosen for our landing was in a small coll beside a parasitic vent on the side of the volcano. At some time lava had forced its way sideways rather than erupt from the main vent far above us. The vent was about 100 metres away and looked rather degraded and with a red oxidised rim. We had little idea how far it was to the top, or what we would find.

Below us was a superb view; we could see jungle stretching as far as the eye could focus. In the distance a short stretch of the river glinted in the sunlight, but generally the river was masked by the trees. Nearer at hand were conical hills covered in vegetation, but with a gradation of colour and texture on their flanks that seemed to reflect changing vegetation with height. These must be old volcanic cones now completely clothed in vegetation. There were also a couple of holes in the canopy, and water was just visible in one; maybe these were crater lakes.

It was fine to view the scene and speculate, but we needed more information before we set off into the unknown. Therefore we returned to the Bus to check our pre-landing view of the area, as we needed to know our elevation, the height of the volcano and our height above the general level of the Carboniferous jungle. We also wondered whether the holes we could see in the forest canopy to the west hid circular lakes that might be filling volcanic craters. Apparently we were 510 metres above local sea level; the volcano rose to 940 metres, and the jungle below graded down to about 100 metres, but there was uneven ground masked by the vegetation. We suspected that lava flows lay beneath the vegetation. There were also three circular lakes within small ring structures, one only about 100 metres across. We fixed co-ordinates for these features, as we would be walking blind in the forest.

Following a council of war it was agreed that slogging to the top of the volcano just to see a crater and a big view was out of the question in the time available. None of us was a keen Munro-bagger, and the call of the jungle was stronger than the climbers' urge to reach the summit. The 400 metres downhill to the jungle was fine, but we had to allow time to climb back up to the Bus before sunset. Our agreed plan then was to visit the jungle, and if the terrain turned out to be reasonable we would try and visit one of the circular lakes.

Going downhill was relatively easy, as the surface was reasonably stabilised. Lichens and moss grew on boulders and clung in crevices. A few small lycopods had managed to take root, but there had not been enough time since the last eruption for a mature plant community to develop. It was not a case of a natural tree line, because other higher but inactive volcanoes were well covered with forest.

We were soon out of sight of the Bus, but the small parasitic vent gave us a marker. The vegetation became thicker, and we forced our way through the young scrub. Crossing gullies required care; the sides were steep and showed layers of tuff and boulders of lava together with lumps that had the contorted and frozen shapes of volcanic bombs. The larger debris was concentrated in the bottoms of the gullies, and slowed our pace as we clambered around large boulders. Water obviously swept down the gullies from time to time, and small trees and ferns clung to crevices, but many had been swept over and uprooted by the force of the torrent.

Progress was slowing down, and the vegetation in the gully was now above head height. We had no line of sight to any familiar features and so had to rely on the co-ordinates of features we had recorded. We had a small hand-held monitor with a screen showing the Bus position, the volcano and the lakes, together with a red moving point that showed our position and left a trail to show the route we had taken. We had a zoom facility that claimed to be accurate to within a few metres. We might well need this for retracing our steps through the jungle.

Steve carried the machete, and left a trail of cut vegetation and blazes on prominent trees. This was our insurance policy in the event of technical equipment failure. The machete was also useful for turning over stones and logs, without getting too close to any hidden wildlife. The need for caution was made clear when a large boulder toppled over in the gully as Steve stepped on it, lost his footing and sat down hard.

'Don't move,' shouted Hans. 'Beside your foot, scorpion—and large.'

Steve raised his head to see two large threatening claws either side of a black shiny body, and a long tail tipped with a fearsome sting waving angrily and looking to defend its owner's patch against all comers. Steve slowly withdrew his foot and stood up, moving a pace to the side. The scorpion turned and continued to threaten this foot that had destroyed its hiding place. The foot and the scorpion were about the same size, so maybe the animal was prepared to fight the foot for territory.

'Well, Steve, are you going to replace its rock?'

'No, I think we'll let it choose a new home all on its own.'

Leaving the scorpion to calm, down we clambered carefully out of the rocky gully on to a ridge. Here progress was easier, and we left a large cut branch on the ridge to mark where we had crossed scorpion gully. Keeping to the ridge we were soon under tree cover, and the undergrowth cleared as the canopy above cut out light to the forest floor. We still had fallen trees and rotting logs to contend with, but at least it was dry and firm under foot and there was no danger of sinking in a swamp. The forest floor was still damp, and vegetation decay seemed to be rapid. There were signs of fungi everywhere, with some rotting fallen trees covered in delicate caps of grey, brown and red toadstools, the reproductive bodies of the mass of fungal filaments that was devouring the fallen tree. Hans was most impressed, and could have spent a whole day examining a single log. These features of decay and recycling of organic matter are very rare in the fossil record.

There were also signs of disturbance of the forest floor: vague trackways up to thirty centimetres wide where animals had ploughed through the plant litter. Amphibians, proto-reptiles or large arthropods were possibilities. The trackways seemed to meander about, so we tried following them, but did not catch up with the perpetrator. Eventually Steve tracked one to a fallen tree. The log was partly rotten and about sixty centimetres in diameter. The trail seemed to disappear into a slot dug under the log.

'Shall we see if it is at home?' suggested Steve.

'Of course, but carefully; it might be an even bigger and angrier scorpion.'

'It will take all three of us to roll the log over, and it will be slimy and messy with all the moss and fungus.'

'I don't mind. Let's do it,' I agreed.

We set to, trying to pull the log and roll it towards us, so that we kept it between ourselves and anything hiding underneath. The soft rotten wood was difficult to grip, but after rocking it back and forth a few times we ripped it clear of the forest floor and could see underneath. The first obvious feature was the mat of forest floor that was attached to the log. White strands and mats of fungal filaments extended from the log into the forest litter, binding leaves and twigs together. There was movement everywhere; centipedes, millipedes, harvestmen spiders and beasts like large woodlice. However, there was a large smooth hollow under the log where the trackway had disappeared.

'Oh, what a pity, it's not at home.'

'It may be. From here I can see that the slot goes farther down under that mat of fungus, and look at those pellets—must be its faeces.'

Steve gently probed with the machete and pulled back the mat of fungus and leaves.

'There is something here I think. I felt movement.'

'Don't cut it. Let's use a stick.'

A probe was prepared, and it soon became apparent that something was at home in this shallow burrow under the log, and it was not too keen to come out. We had to peel back more of the top of the burrow before a head and two rather agitated antennae appeared.

'What is it Steve? You are the arthropod man.'

'Wait till we see it all.'

As if in obedience, the animal began to emerge from its lair, and it just kept on coming, and Steve kept counting the segments. When he reached twenty he announced, 'It's *Arthropleura* or a close relative. Just a harmless vegetarian and detritus feeder, probably comes out at night to scavenge for food that falls from the canopy. It doesn't look as though it would climb trees.'

With that he presented it with a stick to crawl over and lifted the margin of the animal to reveal a sturdy pair of legs on each segment. The whole animal was more than fifty centimetres long and looked like a giant elongated woodlouse. We put the log back in the position we found it, and left the animal to do some home repairs. Hopefully it would go back under the log, and we even helped by clearing the entrance to its lair. Some excellent fossil trackways of these big arthropleurids have been found: one on the island of Arran; and in Nova Scotia some appear to meander about between fossil tree stumps. Body fossils have been found but are very uncommon.

We pressed on through the forest, still gently downhill, and now resisting the temptation to look for beasts under logs, as we had to get a move on. When we checked our co-ordinates it seemed that we had about a kilometre to go to the smallest lake. The land flattened out, but after some 800 metres the topography suddenly changed beneath the trees. Rough angular rock emerged from the ground and we had to clamber up a very rough slope.

'What is this? It can't be very old. There isn't much soil cover.'

'The rock is very black, looks like basalt to me, and I guess we have just climbed up the face of a lava flow,' I suggested.

'According to the monitor we are only a couple of hundred metres from the small lake.'

We could not see the lake, but we could chart the direction on the monitor. So we set off over the rough lava surface on a bearing to take us to the lake. The vegetation was reasonably thin and there were no giant trees here. Maybe the lava was quite recent, and the lava top had not yet been fully reclaimed by the forest.

'Careful! There is a big round hole here.'

'That's amazing, perfectly round and vertical, just like a pothole.'

'That must have been a tree,' said Hans. 'The lava chilled around the tree, which has since fallen, and the trunk has rotted in the hole. Rather like the fossil tree in Tertiary lava preserved at Carsaig on Mull.'

'The tree is not lying here so it must have completely rotted away very quickly and left no trace.'

'No, I think the tree must have been burnt by the heat of the lava. The plants here are all young and must have colonised the bare lava surface.'

We had agreed on the burning when there was sudden noise, a low rumble, then the ground shook, not violently, but perceptibly. Trees moved and rustled, and a few dead fronds and twigs fell to the ground. There was a creaking noise from the jungle followed by the ripping crash and splintering of a fallen forest giant.

'That was an earthquake! Not very strong, but significant in an active volcanic area.'

'Come on, let's find the lake.'

Soon we had crossed the lava, and were walking uphill on reddish soil, with small pieces of lava; this looked like weathered tuff. Then we were standing at the top of a cliff of crudely bedded tuff, looking down into a circular lake about 150 metres across. It was ringed with cliffs, largely draped in vegetation, notably climbers that had scrambled up the cliffs using rocks and tree ferns for support. The calm surface of the lake was dark, reflecting the vegetation. There seemed to be no easy way down, and no evidence of any shallows. It looked like a vent, and we were standing on the eroded rim of a tuff cone. From the rim we could also catch a glimpse of the larger lake only some 500 metres away. It appeared to be at a different level.

There was another rumble, more gentle shaking, and ripples disturbed the lake surface. They moved concentrically from the margins to the centre where they met and then diverged again. It was an interesting sight, but induced a more nervous edge to the group.

'OK, one earthquake is enough for me, I hope we don't see an eruption.'

'The bit of the volcano we landed on did not look active,' I reassured them. 'I can't see an ash cloud yet.'

'Shall we set off back?'

'Well, let's just have a quick look at the other lake. We have plenty of time.'

We had stopped searching for animals, and Hans was not paying so much attention to the plants—fewer notes were being taken. The level of ready banter had also decreased, and we walked more quickly than before. Very soon we were at the rim of the second lake. It was also bordered by cliffs. Some areas were steep, and a lava flow was clearly standing out from the softer bedded tuffs; there were deep fissures in the rock. There was also a distinct low point in the cliffs. This looked much more like a crater lake: brown and red streaking appeared on patches of bare cliff, and the low point in the margin looked as if it might once have been a lava spill point. Over a large area downslope from the spill point the vegetation was a lighter green. Maybe it was underlain by lava from this vent, and the vegetation was at a different stage of development. Again the lake looked deep and dark, and there was no shallow water fringe.

We sat and took stock of the situation. We did not have time to go down to the spill point; we would probably learn little more, and then there was the uphill journey back to the Bus. A clatter from the cliffs to our right grabbed our attention, and a boulder and small stones bounced down into the lake with a series of splashes.

'That cliff looks pretty unstable. Maybe those small quakes triggered the rockfall.'

'Looks like the whole cliff face could fall into the lake at any time. I would not want to be down by that spill point if it falls. We would all be washed down the valley.'

'Excellent, Hans, a health-and-safety reason not to slog down to the lake and back.'

Another rumble hit us, stronger this time, and the earth shook significantly. Immediately more debris started to fall from the cliff face, but it did not stop. The falls continued.

'Get away from the edge, I think the whole edge could go.'

'Quick, the top is moving.'

As we retreated and watched, an arcuate crack opened behind the cliff and with a thunderous roar a section of the cliff plummeted into the lake. A fountain of water shot into the air, and a large wave headed over the lake. We were still watching the wave when there was an explosion. The water surface welled up in a dome and burst; a white cloud remained and subsided to the lake. The water had turned orange around the site of the eruption. Water was now pouring over the spill point, and smashing downhill through the forest. Then the whole lake began to bubble like a glass of fizzy water, but with the rapid change of colour to orange it was more like Irn-Bru. The white cloud was still below us and was being fed from the lake. The cloud flowed over the spill point and down the valley.

'Why are we just standing here? We need to get away fast, and upwind and uphill if possible. That landslide triggered a CO_2 gas release. If we were down there in that cloud we would be dead by now.'

'Is that the end of it ?' Steve asked.

'We'd better hope so,' volunteered Hans.

'Maybe yes, maybe no,' I ventured.

'Very helpful, aren't you.'

'Well, I think this is the critical time. Maybe there is not an enormous quantity of gas in there and that is all there is to release; in which case we are lucky. It does look as though the bubbling is getting less intense.'

'Why the brown colour? It looks like iron.'

'Indeed, it is iron—ferric iron, which is brown like rust. The lake is likely to be 100 metres deep with anoxic bottom waters. CO_2 and methane are dissolved and held in the deep water under pressure, along with iron that is in the ferrous form. When the landslide smashed into the lake it disturbed the pressure balance in the water and had an effect rather like the uncorking of a shaken bottle or a can of fizzy drink. The CO_2 came out of solution as bubbles and rose to the surface. A violent release caused the gas explosion. The red colour is due to the iron turning from ferrous to ferric on contact with oxygen from the atmosphere.'

'Where does the CO_2 come from?'

'In modern lakes that erupt in this way, the CO_2 is virtually all of magmatic origin, part of the volcanic process in the area,' I explained. 'The classic modern examples are the eruptions of Lake Monoun and Lake Nyos in Cameroon. The latter erupted in 1986 and killed over 1,700 people, many of them overcome by CO_2 which is fatal when present at 10% or more. The cloud is heavier than air and flows downhill, filling hollows and killing all animals.'

We had been lucky. If the lake had been deeper, and held more CO_2, we might have been overcome by the gas. We had survived, but many animals would have died, overcome by the gas. Maybe similar events were responsible for the preservation of amphibians and proto-reptiles in the Carboniferous lake at East Kirkton in Lothian. These deposits contain limestone in mounds that may have formed around gas vents in a lake. However, the East Kirkton lake is interpreted as having been shallow, so catastrophic gas release is unlikely. Animals and plants preserved at East Kirkton were probably killed by toxic and/or hot lake waters. The same thing happens to animals that live near the hot springs in Yellowstone

Park: the occasional buffalo falls in, and other animals drown when chased by predators, no doubt sometimes followed by the predator.

We trudged back uphill, following our route to the Bus. The mess we had made of the vegetation on the way down clearly marked our path. We stopped at the *Arthropleura* log for a brief rest, and were pleased to see that our friend had been housekeeping—there was a pile of debris pushed out of the burrow and it had clearly resumed residence. There were also several more earthquakes. Something was clearly going on, and the lake eruption had just been a side-effect caused by the rockfall.

It was with relief that we emerged from the forest on to the flank of the volcano. Looking up the hill into the sun, we could make out the dimple of the small crater beside the landing site. The evening was fine, although a small cloud hung over the summit of the volcano. This was the tiring bit, steep uphill, with packs feeling heavy, but at least there was no shortage of oxygen. Squinting uphill into the sun the cloud looked bigger, and when we all stood still and there was no crunch of boots on ash we could hear noises, dull thuds and explosions. Then the truth dawned; the volcano was erupting. The cloud was ash being ejected from a vent somewhere on the far side of the volcano.

We were only fifteen minutes from the Bus. All effort now went into getting back as fast as possible. It was just ironic that as the ash cloud grew we had to walk directly towards it. Then we all noticed the sizes of the volcanic bombs with more interest. We needed to be out of here before any of that stuff started flying about. Every now and then fine ash fell on us. We stopped to haul out face masks to avoid breathing in the dust that can clog airways and lungs. It was a great relief to find the Bus upright and only dusted with ash.

The Bus rose in a cloud of grey ash and flew around the volcano. A vent on the far side was spewing out ash and lava. The flames were spectacular, and where the lava reached the forest there was an intense forest fire; even green vegetation seemed to be burning with almost explosive vigour. The high level of oxygen in the air was clearly fuelling the fire.

'Good job we didn't try to light a camp fire,' mused Hans.

Excursion 7

The Elgin Reptiles

TIME: Late Permian.

LOCATION: Elgin and Hopeman, Moray.

OBJECTIVES: To visit the great dune system that bordered the Moray Firth basin in the early Permian, and hunt for signs of the Elgin Reptiles.

THE MODERN EVIDENCE: Fossil dune sandstones on the shore and in quarries at Hopeman that contain a variety of fossil reptile footprints, combined with the very scarce fossils of Permian reptiles that have been found in the Elgin area.

The end of the Permian was a tough time to be alive. This was one of the major times of mass extinction in the geological record, a time when a high percentage of species died out. The ecological vacuum created was gradually filled in the Triassic Period, but further extinctions signalled the end of Triassic time, and species diversity did not really recover until extensive shelf seas covered the continental margins in the early Jurassic.

In Britain the Permian and Triassic periods (290–205 million years ago), commonly referred to as the Permo-Trias, are represented by red mudstones, sandstones and conglomerates of the New Red Sandstone. Britain was generally a land area, and marine deposits are restricted to the Permian Magnesian Limestone of northeast England, and a brief sniff of the sea extending into England from the east in the mid Trias—the edge of the Muschelkalk Sea, whose limestone deposits are richly fossiliferous in continental Europe. At the end of Triassic time a marine transgression from the south heralded the general marine shelf conditions of the early Jurassic, and the highland areas of Scotland became islands in a warm Jurassic sea.

Throughout the Permo-Trias Britain lay in the northern arid climatic belt. In simple terms it was drifting north from a position in the southern arid belt during deposition of the Old Red Sandstone; during the Carboniferous it passed through the hot, wet equatorial belt, and entered the northern arid belt in late Carboniferous–early Permian time.

During the Permian, the North Sea Basin area was actively subsiding and, following initial early Permian deposition of wadi sands and gravel on an eroded land surface, extensive dune sands developed. These dune sands are now the sandstone reservoirs of the gas fields offshore from East Anglia in the southern North Sea. This arid basin was well below world sea level, and eventually the sea flooded in, producing an inland sea in a hot arid region. The result was rapid evaporation and concentration of salts, producing a series of salt deposits representing evaporation of the basin to dryness followed by marine flooding. This process was repeated in the North Sea area four or five times. The Magnesian Limestone of northeast England seen on the coast in County Durham is the marginal deposit of this evaporitic basin, the centre of which contains very thick salt deposits—the raw material of the Billingham salt industry. By Triassic times the North Sea was no longer a major salt basin, but local salt basins still formed, as in Cheshire and Somerset, though the latter has not been exploited.

Therefore, at the end of the Permian, much of the present land area of Britain was undergoing erosion, and deposits of fluvial gravels and aeolian dune sands occupied locally subsiding basin areas. The Inner Moray Firth was one such area of subsidence, and the only remaining onshore deposits are found in the Elgin area to the south of the Moray Firth. These onshore deposits consist of sandstones deposited by rivers and wind-blown sand dunes, of which the Hopeman Sandstone is a fine example of a dune sandstone (Fig 7.1). We know from oil exploration drilling results that offshore in the Moray Firth there was a saline lake with sand and gravel around the margins, and muds and silts together with minor salt deposits in the centre. Thus if we were to have visited Elgin in the late Permian we would have seen an elevated hilly area of the Grampians, with rock exposed. River valleys would have carried eroded debris from the hills north towards the Moray Firth, and an east–west belt of large dunes, tens of metres high, would have formed a margin between the land area and the saline lake of the Moray Firth.

It is not easy to find a place on the modern Earth where we could view a comparable scene, in a single glance. The eroding land surface and sand-filled valley systems can be seen at the margins of many desert areas such as the Sahara in northern Africa and the great sand seas of Saudi Arabia. The dunes of Namibia march across country and partially block old river valleys with sand, so disrupting the drainage system. This situation might provide some comparisons with Permian dunes near Hopeman, which probably blocked drainage channels leading from the Grampian uplands to the Moray Firth Basin. The salt lake of

Lake Eyre in central Australia has dunes fringing the shoreline and lies within a subsiding basin, now below world sea level. Hence Lake Eyre provides an analogy for the Hopeman dunes fringing a salt lake in the Moray Firth.

However, it is seldom possible to find a perfect analogue for the ancient in the modern world. Whilst we can match the physical parameters of latitude, temperature and rainfall, we cannot recreate a Permian flora. We do not know how well adapted the flora was to arid conditions, how it affected erosion rates by binding the surface deposits and how well the soils might have retained moisture.

Thus, in view of interest in the desertification process that is so badly affecting the desert margins of the present day, and reducing capacity for desperately needed local food production, the expansion and contraction of ancient deserts is a topic of great interest. The Permo-Trias is an obvious target since there was significant vegetation that could be monitored remotely in terms of expansion and contraction of vegetated areas. It also corresponded with a major period of expanding and contracting ice caps, and consequent changes in world ocean level. Thus there are similarities with the present world situation where we are enjoying a (probably) brief interglacial, although the main concern in society is climatic warming due to the activities of man. We also know that recent glacial episodes had a great influence on climate in present-day arid belts, causing expansion and contraction of lakes such as Lake Eyre in Australia and Lake Chad in west Africa. Contraction and squeezing of the trade-wind belts closer to the equator by expanding ice caps greatly increases wind speed and sand transport by the wind. This results in larger sand grains being transported, to form bigger dunes than form at the present.

We can see the relics of these stronger winds today. In Saudi Arabia, the Sahara and Namibia many of the largest dunes are in effect fossil dunes. They are not moving at present. They are made of coarser sand than the recent dunes that merely modify these giant relics of stronger wind regimes of the last ice age. The grain size of wind-blown sand gives a measure of the maximum wind speed. The well-rounded two-millimetre grains of ancient dunes imply frequent wind speeds greater than those experienced in these regions at the present. Similarly, dune systems that cover vast areas of Australia are stabilised by vegetation and are also fossil, some over 100,000 years in age. Bearing these features in mind, we therefore set off in the Bus to visit the southern shores of the Inner Moray Firth, somewhere near Hopeman, in the late Permian.

Before any safari it is as well to review the fauna and flora of the area. It is not advisable to wander into the African bush unaware of the possible danger from

Fig 7.1 Permian Hopeman Sandstone with large-scale dune cross-bedding in cliffs near Hopeman. (Photograph courtesy of Ken Glennie)

lions, elephant or water buffalo, and in the Rockies of North America all visitors are instructed to take precautions against meeting a bear, although everybody hopes to see one at a safe distance.

In Australia things are always different. If wandering in the true outback there are no large predators to beware of, but everyone is wary of poisonous spiders and snakes. The heart always jumps when a stick bounces up from a careless footstep and hits the legs; was that a snake? Will I die? The Australian exception is the interaction of humans and saltwater crocodiles in the north, where occasional fatalities occur, a large proportion of which are alcohol-related, as was the case with the swimmer who thought the croc would give him a tow. (He had ignored all the 'Crocodiles are present in these waters' and the 'No swimming and fishing' notices and was in effect signing his own death warrant. His body was a free lunch to the grateful reptile.) He certainly deserved a 'Darwin Award'.

So what might be expected on a journey into the late Permian of the Elgin area? On the animal side there were certainly reptiles—the fossil fauna commonly referred to as the 'Elgin Reptiles'. These fossils have been collected and studied since the first half of the nineteenth century. In the early days there was much debate over the age of the fauna, but we now know that reptile fossils of two distinct ages are found in the Elgin area. The older fauna is late Permian–early Triassic in age, and the original material was collected from Cutties Hillock, the first finds being made in 1885. The younger fauna is late Triassic in age and most specimens came from the Lossiemouth area. The geology of the area was difficult to elucidate for the early workers, since the sandstones of the Upper Old Red Sandstone (Devonian age) and those of the New Red Sandstone (Permo-Triassic age) in the area are very similar, and can even occur in the same quarry. This is despite the fact that there is an age gap of about 100 million years between them, including all of the Carboniferous Period. Thus Devonian fossil fish can be found in the lower strata in the quarry, and Permian reptiles in the sandstone above. A further complication arose because the bony body scales of one of the late Triassic reptiles, *Stagonolepis* by name, were originally identified as large fish scales and thought to be Devonian in age.

Once the complexity of the geology is sorted out, we seem to be left with the fossil remains of some four types of late Permian reptiles, three of them closely related. However, this is likely to represent only a (very) small sample of the animals that lived in the area. The fossils are rare, and represent preservation of individuals that happened to die in circumstances in which they could be preserved as fossils. In many cases the bones are found in sandstones that were deposited as sand dunes, or in sandstones from the base of a fossil sand dune. It appears that these animals perished in an area of shifting dunes, and before the bones could disintegrate in the atmosphere they were covered by the advancing sand dunes, and hence preserved. So these were maybe the unlucky animals that expired in unusual circumstances in the dunes. Did they have to cross the dune field to find water? Were they on migration, or were they just old and lost? These are questions we cannot answer, but it is unlikely that they were in their main feeding area, since there is little for vegetarian reptiles to eat in an area of mobile sand dunes.

The actual fossils of these reptiles are frequently unusual in that no bone material may remain, only a hole in the sandstone from which all bone has been dissolved by fluids percolating through the pores in the sandstone. In order to study the 'bones' the technique used to be to pour rubber solution into the holes and allow it to set—then pulling or breaking out the rubber replica of the fossil. In 1997 a large block of stone from Clashach Quarry near Hopeman had to be split because it was too heavy for the lorry sent to collect it. On splitting the rock, a hole was noticed on the cut face. To the quarryman this was a minor disaster, indicating that the stone was flawed and had to be rejected. However, this was fantastically good news for the palaeontologist; the hole clearly represented bone that had been dissolved out of the rock. After cutting to a manageable size, the block was studied using x-ray techniques and a medical scanner. A computer-

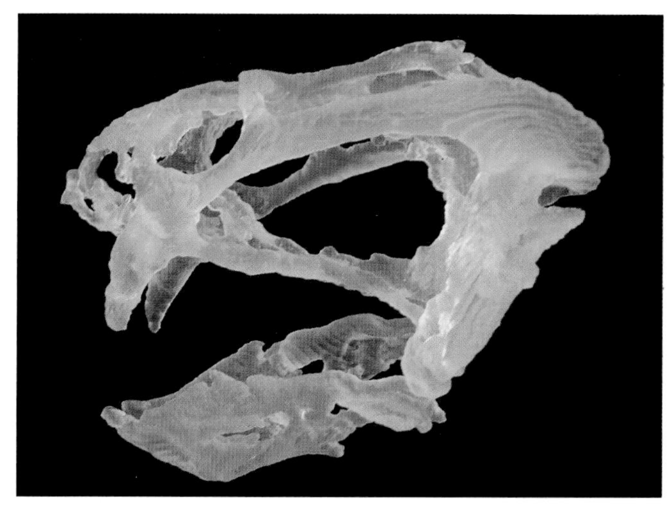

Fig 7.2 Skull of the late Permian reptile *Dicynodon*. The 3D model is derived from an MRI scan of a mould of the skull in a block of sandstone from Clashach Quarry, Hopeman. The skull is *c.*23 cm long.

generated 3D image of the 'hole' was produced to reveal a skull of the reptile *Dicynodon*. The images produced showed excellent anatomical detail, there being no need to break open the block containing the cavity (Fig 7.2).

In many desert areas, and in coastal dune fields, shifting sand covers carcasses and illustrates the potential for preservation. If no net deposition takes place as a dune migrates over the carcass, it may re-emerge again—as has happened in the case of camels and victims of the Second World War in the deserts of north Africa. If, however, subsidence is taking place, and sand is accumulating to fill the space created by subsidence, than the carcass may never be re-exposed, and is at the start of its burial journey to becoming a fossil.

In any event, something unusual has to happen to the carcass for it to be preserved; the normal situation for a land-living animal is that it will not join the fossil record. Just consider the teeming herds of gnu and zebra on the Serengeti plains of Africa. If the bones of all the animals that died, or were eaten by predators, were to remain on the surface, the plains would by now be knee deep in bones. That is not the case. The killer eats his fill, and leaves the rest to the scavengers. The bones break down slowly in the atmosphere, and nothing is left for the geological record. It is only, say, on the shores of a subsiding rift valley lake, where the bones can be buried in accumulating sediment, that there is a chance of preservation. In fact a large chapter of our knowledge on the early history and evolution of man comes from ancient lakeshore deposits in the rift valley of east Africa, famously documented by the Leakey family.

So our Permian quartet of reptiles—*Geikia*, *Gordonia*, *Dicynodon* and *Elginia*—are likely to be a poor representation of the fauna of the time. The first three named are dicynodonts and were generally similar in form, and appear singularly unattractive in reconstruction. Imagine a rather squat, fat, hairless dog with stout legs and no tail worth the notice. In the upper jaw dicynodonts possessed two tusks, showing differentiation of the teeth as seen in the later mammals. Hence these reptiles are known as the 'mammal-like reptiles'. The general dentition of the animals indicates that they were vegetarians, chewing up coarse plant material. The tusks, which are variably developed in the different animals, might have been used for digging up roots, as do pigs. They might also have been helpful in fighting or defence, like a boar. Another possibility is that the animals could excavate burrows, a useful ability in a world where water may have been in short supply. Many animals will dig to keep open water holes in the dry season, and maybe these animals did the same. Possibly the land iguanas of the Galapagos Islands provide an analogue of reptiles living in an arid area. These animals make burrows; Charles Darwin recorded that on one island it was difficult to find a flat area free of land iguana burrows on which to pitch a tent.

The last of our quartet, *Elginia*, was a cotylosaur reptile and is known from fragmentary material; the skull is extraordinary, bearing sixteen robust spines, the largest pair bearing a resemblance to a cow's horns (Fig 7.3). The teeth were small, and again it was a vegetarian. The spiny skull would have been useful in defence, but maybe it also functioned in displays of seniority in the mating season. Like the dicynodonts, *Elginia* was also a rather small, squat animal with a short tail.

So our animal gallery is somewhat disappointing in diversity, and only reveals a fauna of plant eaters. It would be most unusual if there were no predators around at the time; but top predators are usually a minor numerical proportion of the population. There must be more fossils to be found in the Elgin area. We have the evidence for the existence of other animals in the form of trackways preserved in the dune sandstones (Fig 7.4). In recent years many have been found in Clashach Quarry on the coast near Hopeman. The most common type was probably produced by dicynodonts, but rare examples showing a tail drag mark were made by an animal that has not yet been found as a fossil skeleton.

The remarkable preservation of the rare reptiles is good fortune, but we have no such luck when it comes to the plants. A review of Permian and Triassic plant fossils from the Elgin area is quick and simple—there are none. Plant material is subject to rapid decay and oxidation, and hence is rarely found fossilised in sandstones deposited on a land surface. Most plant fossils originate from a leaf or twig being buried under anoxic conditions, where the lack of oxygen inhibits breakdown of plant cuticle. Such conditions occur in stagnant lakes, swamps

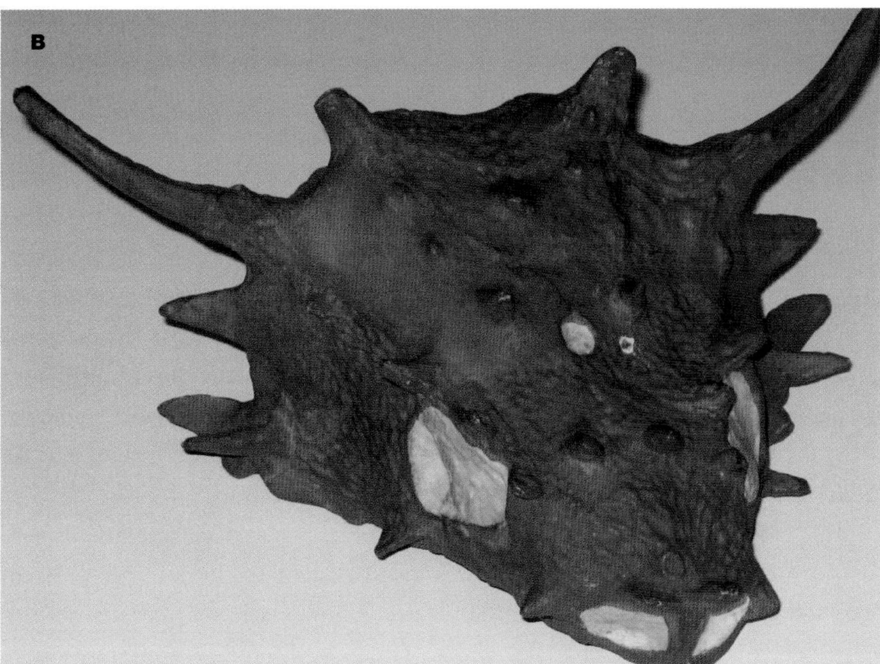

Fig 7.3 **A:** Model of the Permian reptile *Elginia* in the Elgin Museum. **B:** Cast of the fossil skull. The model, at natural size, is 85 cm long, while the skull is 13 cm long. (Model by Stephen Caine; photographs courtesy of Bill Dalgarno.)

Fig 7.4 Permian reptile trackway in Hopeman Sandstone. Clashach Quarry, Hopeman. Scale bar 10 cm. (Photograph courtesy of Ken Glennie.)

and marshes, and some nearshore marine mud deposits. This gives a strong bias to lowland floras in the geological record. Plants must have been present during the Permian in the Elgin area for use as reptile fodder. Terrestrial vertebrates are seldom found far from a source of food, even when migration is involved. So if there are no local plants preserved, what might have been growing at the time?

Since we have no direct fossil evidence of the Permian vegetation from the Elgin area, we have to rely on general information gleaned from farther afield. We do know that floras changed radically from those so well preserved in Coal Measures of the Carboniferous Period, but that was before the late Permian, and Britain had drifted from the wet equatorial zone northwards into the northern arid belt. The lush forests of *Calamites* and *Lepidodendron* had died out in Britain, even though forest relics remained in some parts of the world, restricted to suitable environments.

By the time that the reptiles were roaming the Elgin area the Variscan Orogeny had come and gone, uplifting the Variscan Highlands to the south in Brittany and southwest England. Such highlands undoubtedly affected climate, and the Elgin area may have lain in a rain shadow to the north of the mountains.

In the drier conditions of the late Permian, plants had to be able to survive on less water, and maybe withstand periods of drought. Under this regime the conifers became dominant and were accompanied by cycads, and various types of ferns. These were probably the tougher elements of the flora that formed the diet of the vegetarian Elgin reptiles.

In the Elgin area there were sand dunes, and the rocks contain no evidence that the dunes were vegetated. We do, however, find a lot of burrows (Fig 7.5) in the sandstone, indicating that there was life and therefore some dampness, maybe from overnight dew. To the south of the dune belt the ancient eroded bedrock of the Grampian area was exposed as hills with valleys draining north to the Moray Firth. The rivers may have flowed seasonally, being reduced to a series of disconnected ponds in the dry season. As is common in desert areas, rare storms may have produced flash floods that roared down the valleys from the Grampian Highlands. Valley areas would have been oases of rich plant growth, but plants probably clung to rocky crevices in the hillier areas, eeking out a precarious living on water seeping from the rock.

The vegetation of the low-lying subtropical areas of Western Australia provides a modern example of vegetation adapted to dry conditions, where small bushes have very deep roots extending many metres to the underlying water table. The unique nature of the west Australian flora with banksias and grass trees has evolved over millions of years, and there was plenty of time for the plants of the Permian to cope with the changing climatic conditions of those times. Similarly there was time for the vegetarian reptiles to adapt to the available salad. In essence the fauna and flora evolved in unison, continuously developing new strategies for survival. Without the fossils we can only indulge in speculation, and what use is that when we see the amazing interdependence of plants and animals in the modern world? How could holes in the thorns of a 'fossil' acacia tree ever be interpreted as the homes of ants that defended the tree, and hence their homes, from attack? Truth observed in the modern world is far more bizarre than the speculations of palaeontologists.

Having reviewed the known 'facts' relating to environment, flora and fauna in the late Permian of the Elgin area, it is now time to stretch the imagination and take a stroll from the shores of the saline lake that occupied the Moray Firth, south through the dune field and into a valley draining from the Grampian uplands in the late Permian. My companion on this excursion was to be Carol, who has made a detailed study of Permian reptile tracks and has discovered many new forms from the Hopeman area.

After a frustrating few hours of time search on board the Bus, we located reasonable palaeogeography for the late Permian. The absolute dating of this part of the geological column in Scotland is not well defined, so we had to track back

Fig 7.5 Burrows, probably made by small arthropods, preserved in Hopeman Sandstone from Clashach Quarry, Hopeman.

and forth in time until we had a view from the Bus that incorporated a shimmering saline lake flat bordered by a strip of sand sea. The large dunes threw shadows in elegant curves across the slip faces in the pink light of early dawn. Many of the large dune faces were directed to the west, indicating winds from the east, but there were clear local variations caused by the topography. To the south of the dune belt a more jagged and varicoloured landscape of rocky hills and valleys, still part shrouded in morning mist, faded south into distant greys like a Colin Baxter photograph. Drainage was to the north, but the valleys seemed to be truncated and blocked by the dunes that bordered the lake margin. It was also apparent that the lake margin was remarkably straight. There had to be an underlying explanation: maybe this was the faulted basin margin forming a sharp dividing line between the saline basin deposits and the marginal dunes.

The Bus bumped to a halt on a hard surface—so far so good. We were on target. The view to the north was a monotonous spread of sand with glistening white in the hollows, shallow abandoned channels testifying to the intermittent presence of water in quantity, but now the scene appeared to be dry and parched. To the south was a much more interesting scene: only a few hundred metres away, elegant dunes formed a barrier to our view of the land beyond. It was time to go and investigate.

Hot! A baking wind was blowing along the shore, and sand grains bounced along the surface with the characteristic gentle hissing sound of blowing sand. The surface was crusty with loose dry sand underneath, and the sand crust cracked in thin slabs away from our footprints. The surface crust was only a few millimetres thick, the sand grains being stuck together with salt that had been deposited from evaporating water at the surface. Not far away the surface was stony: a pavement of small pebbles up to about ten centimetres in size, with the tops of the pebbles smooth, fluted to smooth faces with sharp edges and glistening with polish from the regular sandblasting they received (Fig 7.6). However, picking them up revealed the undersides to have the rounded and pitted surface characteristic of water transport. Clearly the pebbles had arrived in this spot by water transport, but had then been blasted by the wind into the form of ventifacts. The wind could not move the pebbles, so they remained as a protective pavement, preventing further wind erosion.

How long might this process have taken? In Scotland, in coastal dune areas, ventifacts have formed since the last glaciation and the relative stabilisation of the coastline—a period of up to 10,000 years—and occur on surfaces with Mesolithic flint tools. In the Permian the process was probably faster, since the grains being

Fig 7.6 A ventifact (pebble sculpted by the abrasive action of wind-blown sand), 18 cm long, from the pebble pavement below the coastal dunes at Newburgh, Aberdeenshire.

transported were larger, indicating significantly higher wind speeds. But there was still a puzzle. In front of us lay a seemingly continuous line of dunes, so how did water transport the material from the higher land beyond the dunes? Had they been moved before the dunes formed? Were there channels that brought water through the dunes during floods?

Trudging towards the dunes the surface changed to sand forming low-angled, gently rolling surfaces, in places smooth, in others with sweeping patterns of delicate wind ripples. The surface was damp yet firm, but again there was a cohesive crust in places, particularly in areas still shadowed from the morning sun. Dew was still on the ground. We had seen fog in the valleys before we landed, so there was certainly water available. With the sun rising, the temperature climbing and the wind increasing, the dampness could not last long.

As we headed for an impressive dune slip face some 20–25 metres high, the wind was already blowing a plume of sand over the crest of the dune, and depositing its load on the slip face. The face eventually over-steepened, and collapsed with a sheet of sand flowing down the face and encroaching on the rippled sand at the foot of the slope. Thus the dunes gradually migrated downwind.

'I'm not climbing straight up that,' observed Carol.

'Nor me,' I agreed. 'It would be one step forward and two back through soft sand on a steep slope. We would set off avalanches all the way, but it would be great to slide down on the way back.'

We traversed along the base of the dune where wind ripples interfingered with the avalanches of sand from the dune face. The sand was coarse grained, with well-rounded grains up to a millimetre in diameter. Close inspection showed they had frosted surfaces, as seen on a glass bottle that has been sandblasted by the wind at the top of a beach. The slip face of the dune was reducing in height as we approached the end of this particular dune. However, it was clear that there were further dunes for us to pick our way around or over before we reached the rocky outcrops to the south of the dune belt that we had seen from the Bus.

Rounding the end of the dunes we were stopped dead in our tracks by the tracks of another. In front of us was a double line of footprints of rounded form and each with the evidence that the perpetrator had five toes with claws on each foot. Closer examination showed that the front feet were smaller than the back feet, and that as it walked our animal placed its hind foot so close to the previous front foot impression that the two frequently merged. Our eyes followed the tracks into the distance, where they stopped at the foot of a dune face.

'Where did it go?'

'Not into the dune but over it, I guess. The tracks on the dune face must have already been covered by blown sand.'

'How long ago? Are we hot on the trail?'

'Difficult to tell, but some of these prints have been modified by wind scour, and the wind isn't moving the sand here at present, so I guess it was not made this morning. Maybe yesterday, but it can't be more than a few days old,' suggested Carol.

It was an eerie feeling looking at the track of a Permian reptile. How far away is it? Will we meet it? Will it be friendly? These and other questions leapt to mind, and we almost expected to find life around the next dune.

'Here's another track,' announced Carol.

'Not so well preserved, but going in roughly the same direction. It certainly looks as if there are a few animals about.'

We soon arrived at a damp hollow between dunes, which had been blasted out by wind channelled by the dunes. The surface was pebbly in places, representing the base of the dune system, the surface over which the dunes migrated as the wind drove them forwards.

'What's that white patch?'

A curious area of white material stood out from the dark pebbles in the hollow. As we approached, the unison cry was 'bones!'

There, scattered on the surface, were the whitened and disintegrating bones of a reptile. Most bones were reduced to white little slivers by the heat in the sun, and the skeleton was scattered and incomplete: a few ribs and vertebrae, fragments of short robust limb bones, but no sign of a skull.

Had the skeleton been dismembered by a predator or a scavenger? Had the animal died on a sand dune, and the bones been let down on to this surface as the dune migrated? Both scenarios were possible, but we had no time for palaeoforensics, and the bones were probably too decomposed to reveal the telltale bite marks of a predator or scavenger. However, one other thing was apparent: small piles of sand each about the size of a pea dotted the sandy areas between the pebbles. These were the ends of burrows, and a quick examination produced a rather cross little beetle-like animal from one.

'What does a beetle want with a burrow, and what does it eat?' I asked.

The beetle promptly disappeared from sight into the sand without offering a solution. Most probably the beetles came out of the burrows at night in cool conditions to scavenge for food on the dune surface. By day they buried themselves in the cool sand to conserve moisture. (There is a beetle that lives in the deserts of

Namibia that collects water by allowing mist to condense on its body; it stands in such a position that the condensing water trickles to its mouth.) There was also a surprising amount to eat in the dune area: both plant and animal life had been blown into the dune system, and the wind concentrated this lighter material in quiet areas where it was available as food to dune-dwelling animals.

Of course, if there was enough organic material to sustain a population of beetles, there would surely be animals that preyed on the beetles. We might expect to find small lizard-like reptiles to perform this task. Again, the lizards might be nocturnal, and hunt beetles that are on the surface, or they could dig for them in their burrows. Another possible predator to look out for was the scorpion. Scorpions also tend to hide in burrows or under rocks by day, and hunt at night. We were unlikely to meet a scorpion walking about in broad daylight, but we might see scorpion tracks that were quite distinctive, with leg imprints as oblique groups of three, and sometimes with a central drag mark. Such trackways have been found that are 400 million years old, so scorpions are a very long-lived group of animals.

Leaving the beetles in peace we followed the low areas between the dunes, attempting to weave our way through the sandhills. The wind was still rising, and avalanches of sand regularly slid down the dune faces. There were more examples of reptile tracks, some of which appeared to start from nothing and fade out without explanation. Closer inspection revealed that trackways could start and end with erosion wiping the evidence away, or by blown sand covering the tracks. Thus at the present time, when blocks of stone are split in Clashach Quarry, some fossil trackways appear to end without reason because the rock has split along an erosion surface, rather than along the surface on which the animal walked.

We trudged on through the dunes, trying to avoid having to climb the soft dune faces by sticking to the rolling areas covered in wind ripples. The temperature was still rising and now there were no shadows and no shade in the dune hollows. All vestiges of dew had evaporated, and the sand surface was hot and getting hotter. Not being able to see any distant horizon we used, just for fun, a simple compass set to a late Permian pole position for orientation, keeping our modern autotracking device in reserve in case we lost our way in the dunes.

Soon the dunes closed in around us and there was no option but to struggle to a higher level for a more distant view of the surroundings. One passage obliquely up a dune face set off continuous avalanches of sand, leaving an ugly scar on the previously pristine dune face. Sand was blowing in an elegant plume from the dune crest, and as we climbed we were showered with blown sand. Eventually, covered in sand, we reached the dune crest. Every available crevice in clothes and equipment was now gritty with sand. Once over the crest we were in clear air and could breathe and open our eyes, blinking out the sand and dust.

The windward surface of the dune was harder and compacted, easier to walk on and covered with coarse sand ripples. Looking about us, rather than at the sand at our feet, we could see dunes to the east and west, and to the north the shimmering white surface of the playa lake with dancing mirages of non-existent water. To the south lay the hills beyond the dunes, in the distance some dark and jagged, others paler and more rounded. The topography reflected the bedrock geology of the basement with darker fractured metamorphic schists and more rounded features of granite intrusions, but in the foreground we could see ridges of red sandstones, cut by valleys.

'That is the Old Red Sandstone, isn't it?'

'Yes, it looks well exposed,' I agreed. 'It's very tempting to go and look for Devonian fossil fish, but we'd better stick to the job in hand.'

We took a bearing on a valley and set off to find the point where the valley, which presumably carried water in times of flood, met the dune belt. After an hour traversing dunes and trying to stick to the easier walking on the windward slopes, we eventually crested a dune ridge and our objective lay before us, and what a surprise it presented (Fig 7.7).

There was water! There was vegetation! At the mouth of the gorge-like valley lay a pool of clear water maybe fifty metres long and twenty metres wide, with a pebbly shoreline on the far side. A dune face descended below us to the pool's edge—we could have slid almost directly into the water down the dune face. The brilliant patches of green amongst the red rocks and yellow sand made the scene reminiscent of parts of Australia—the pools in Windjana Gorge and Geikie Gorge at the edge of the Kimberley Plateau in the north of Western Australia—and especially the pools in the Murchison Gorge upstream from Kalbarri in Western Australia, where gum trees cling to the red sandstones around pools that are virtually stagnant in the dry season when the river has little flow; but very occasionally they have to withstand floods that roar down the gorge threatening to rip them from the rocky ledges along the gorge.

However, instead of gum trees were a fine display of conifers, cycads and tree ferns clinging to the edge of the river channel and growing from crevices in the red sandstones of the valley sides. There was also a general background green of smaller fern-like vegetation clothing areas where water seeped from the sandstone.

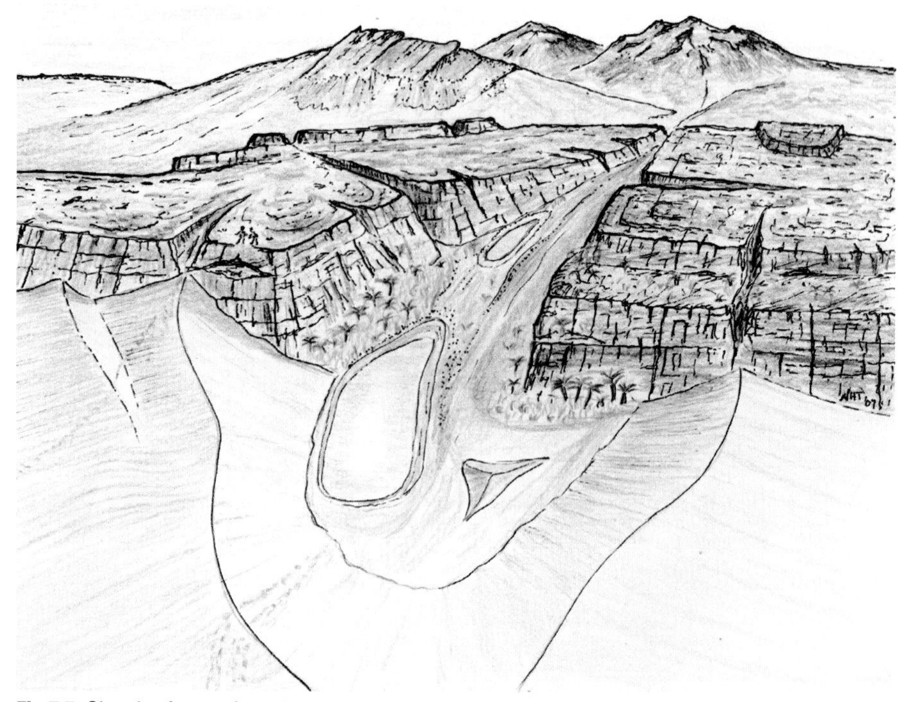

Fig 7.7 Sketch of part of the Excursion 7 route with late Permian sand dunes banked against a cliff of Old Red Sandstone with hills of granite and metamorphic basement in the distance. Water seeping from the sandstone provides pools in the valley and an oasis of vegetation for Permian reptiles.

Now it was apparent why water was present. The Devonian sandstones acted as an aquifer, carrying water from the highlands to the south, and this water seeped out as springs in the valley floor which appeared to have cut down to the water table at the edge of the dunes. Such features are seen in desert areas today, for example in Oman where mountain valleys descend towards the sea and water seeps from the bases of alluvial fans. Habitation and associated agriculture have utilised this environment and its water source for centuries. In Hopeman, in the Permian, we could see the same type of situation, but here the water utilisation was by the biota of the area. We had water; we could see plants; we had seen reptile footprints and bones. Surely we could not be too far from meeting the 'Elgin Reptiles'. Exercising caution we stood and watched, intent for any sound or movement in the scene before us, but no animals showed.

'This is a good waterhole, surely there must be animals,' remarked Carol.

'Agreed, but I can't see anything moving.'

'Look at the beach on the far side. I think the sand is trampled. There are just so many footprints that it is like Glastonbury after the festival.'

'Maybe this is the wrong time of day. It's hot, and animals tend to rest in the shade in the heat of the day. We should be here at dawn or dusk.'

We passed the one pair of binoculars between us, each scanning the scene for signs of life. We felt like a couple of twitchers on a birding expedition, but there was no possibility of birds, and the tick list for all live animals seen so far remained disturbingly low—at one, a beetle. It would be good to be forewarned of possible encounters with larger animals.

There was an increasing nervousness in the exchanges between Carol and myself. Descending to the pool and walking into the valley would give us another viewpoint, although we would be hemmed in by water, dunes and steep valley sides. If anything was down there, we could be meeting it at close quarters rather than safely in a distant view.

The next bit was fun—sliding down the dune face and setting off grain flow avalanches of sand as we slid to the base of the dune and watched the grain flows freeze at the foot of the slope where they overlay coarser wind-rippled sand. As there was still no animal activity, we walked with caution along the sand and pebbles of the narrow strip that lay between the dune face and the water. If there was anything large and angry ahead and it chose to charge at this point, we had two options for avoidance—up the dune face (slow and hard work) or into the water (wet, and what else was in there?).

However, nothing stirred and we plodded to the end of the lake where the valley emerged from the gorge, and there was significant vegetation providing shade and cover. There was hardly any point in discussing the fact that animals had been here—many of them, and not long before. The sand was totally trampled, and it was impossible to make out individual trackways. It was easy, and somewhat reassuring, to see that the sand was littered with the droppings of animals, of about the size a large dog might produce, but clearly composed of fibrous vegetable matter. These appeared to be the cowpats of the Permian, dried and disintegrated to various degrees, but some still fresh and retaining moisture, clearly not more than a day old on the basis of aroma.

Life is full of choices, and we had another to make; to go and look in the vegetation of cycads and ferns, or to stick to the open sand and take a quick walk into the confines of the narrowing valley. We took a compromise, employing basic tracking skills. Skirting the vegetation, it was apparent that sandy paths, or more correctly tunnels, disappeared into the vegetation. These were regularly used trampled paths, made by animals probably less than a metre high. Telltale evidence was provided by old spider webs across the trails above sixty centimetres high. This was a smaller-scale example of a feature seen in the Australian bush. When following a kangaroo trail, a tracker's face continuously bounces off large spider webs strung between the bushes above kangaroo-head height, but on a popular bush walk the webs will already have been broken so this does not occur.

Having ascertained that the Permian scrub probably contained reasonably small vegetarians because of the height of the spider webs across the track, we felt easier, but still carried on up the valley without going into the vegetation like a pair of grouse beaters sent to flush the game.

'Rabbits?' queried Carol.

She was looking at a burrow entrance that was about the size of a rabbit burrow, and had a spread of excavated material around the entrance. It was not the only one; more were scattered through the ferns.

'Either there are several individual burrows or they are connected into some reptilian warren. Luckily we don't have a spade to dig one out, or a terrier to flush one out.'

'We would be on a charge for violation of ancient animal rights. It's probably just as illegal to dig in a Permian reptile burrow as a badger sett. All the same, the burrow shape would be interesting. In the Permian Karroo basin in South Africa some small dicynodont reptiles have been found preserved within spiral burrows. Possibly they died in the burrows or were trapped by floods and drowned.'

The burrows we were looking at were not in the safest situation. They were in the gorge, but dug in the highest available sand. Maybe floodwater seldom reached this height and they were only in danger from flash floods which might occur infrequently, say every ten or 100 years. This was maybe an ancient analogy for the human practice of building houses on floodplains (the clue is in the name!). A house may not flood for years, maybe a lifetime, but eventually a spell of extreme weather will cause the exceptional flood, with inevitable consequences.

'Why do you think they bothered to burrow?' asked Carol.

'My guess would be that it is for protection from predators and night-time cold. Reptiles need to warm up to be fully active, frequently by basking in the sun. At night it would be cold here and they might retain more heat in a burrow. They would also be protected from predators and have a bolt hole if attacked. Imagine if they were out in the cold at night; at dawn they would be sluggish and easy prey for any reptile that had a fast-start mechanism.'

'Like *Dimetrodon*, I suppose. They had enormous sails on their backs, which were well supplied with blood vessels, and could have used them to absorb the early morning heat from the sun, warm up fast and be first to breakfast on the sluggish vegetarians.'

'*Dimetrodon* and the like are not known from Europe as yet, but there could have been similarly adapted predators. We just do not have a good enough fossil record for this time.'

The valley sides closed in, and we were in a typical sand-based channel that only contained flowing water at the times of flood. The bed was dry, and where mud had been deposited in drying pools it had cracked and curled in the sun. Dead, wind-blown vegetation lay in hollows, and a variety of small trails crossed the mud and sand. Small burrow entrances were common; clearly there was a healthy population of invertebrates, and probably small lizard-like reptiles that also had burrows.

Rather than follow the main channel we entered a narrow side channel, only a few metres wide and with rock walls to which ferns and the occasional cycad clung, more commonly on the wall shaded to the morning sun. There was no vegetation on the lower metre of the rock walls; they were well scoured and undercut by the floods that occasionally swept down this mini-canyon. It looked as though rains were infrequent, but storms could result in powerful floods.

Suddenly we spotted an animal, which just stood there, staring at us and wondering what we were and whether we ate small dicynodonts. We stood still and wondered whether the beast used its tusks in an aggressive manner. The floor of this narrow side gorge was sandy and well trampled, and we were on a well-used path.

'Look beyond it—in the shadow,' urged Carol.

'I guess we are outnumbered.'

We seemed to have met the guard; behind him or her there were others lying in the shadows. A few heads were raised as our new acquaintance yawned widely and then snapped its jaws together with some force, making a grunt that immediately produced action from the rest of the tribe. About a dozen individuals edged forward behind the leader.

'I think we have the family.'

'They don't seem too welcoming, despite the size—some are small and the tusks are hardly developed at all. They can't be called attractive; I don't want one as a pet.'

The standoff continued. Our family group of dicynodonts—for that was undoubtedly what they were—shuffled closer together, arranging themselves with the larger individuals at the front, forming a protective shield to the juveniles. The largest was about a metre long, with a fat squat body and short sturdy legs, somewhat splayed from the body to give an aggressive appearance. The jaws ended in a kind of beak and a pair of short tusks protruded from the upper jaws. The skin looked warty and appeared to be brownish with diffuse vertical stripes of dark green. However, they were dirty, and needed a wash to reveal the true colour.

'They see us as a threat; they are clearly on the defensive,' I muttered.

'Yes, I suppose we are rather tall for the Permian.'

'That could mean they have met tall animals and don't like them, as if there may be something hereabouts that is large and bipedal and eats them. They certainly seem to expect an attack.'

'I don't think there was anything tall, fierce and bipedal at this time. I couldn't find anything large and fearsome from the Permian in this part of the world in Mike Benton's book on fossil vertebrates.'

The yawning and jaw snapping increased; it was clearly a threat rather than an indication that they were sleepy.

'I think we should retreat slowly and leave them alone. They may be vegetarians, but I would not like a bite on the ankle from those jaws.'

'Agreed. Let's just back off slowly, and maybe we could go up on to a sandstone ledge to get out of their sight.'

We backed off and their tone changed; they clearly felt that they had won a victory, and they stood their ground. A convenient joint in the sandstone allowed us to haul up on to a ledge some two metres above the sandy floor of the gully. Now out of sight of our dicynodont family, we sat down to review the situation. Time had flown, and it was about time to return to the Bus. The gorge side was not steep, and convenient ledges of sandstone formed a reasonable staircase to the top of the gorge. We could go up over the ledges with their loose rocks and plants clinging in crevices, or we could go back down to the valley floor and return the way we had come. Going up the gorge side was probably easier than trudging back up the large dune face, even though we would have to find a way down the sandstone cliffs to the dunes. Anyway, it was not far, and there would be time to backtrack if required.

'We should have looked more closely at the cliffs at the end of the gorge.'

'Yes, but I think we would have noticed if they looked impossible.'

'OK—go for it.'

The route up was full of surprises: delicate ferns and cycads growing in shady crevices, and mosses in the dampness of small caves etched in the softer sandstone beds. At one point we crouched down at the edge of a deep overhang, to enjoy a bit of shade and have a close-up view of some mosses and small, spore-bearing plants. I moved a rock out of the way, tipping it towards me, and Carol let out a sharp exclamation and recoiled from the rock. I immediately let go of the rock, which fell back into place.

'What was there?' I asked.

'Scorpion.'

'Excellent, how big?'

Just as I went to lift the stone again, more carefully, there was a noise from the darkness under the ledge.

'Something moved in there.'

So here we were, exposed on a gorge side, at a cave mouth, not able to see the back of the cave—not a good defensive position.

Hardly giving us time to retreat from the opening, two sandy coloured animals each the size of a small Border collie, but far less graceful, shot past us and bounded up the ledge.

'What were they, apart from annoyed? Permian mad dog? Did you see the teeth? Certainly not herbivores. I guess we disturbed their rest.'

The cave entrance was narrow, a metre high at the front, but from the sandy ridge at the entrance the floor sloped down into a larger space. The cave was clearly the den of Permian 'dogs'.

'They were scared of us as well. Why? They have never seen a mammal—I still worry that something nasty and as big as us exists here.'

'We should have been more careful. That is just the place where animals lie up during the day. I'm glad they were not bigger, and were scared of us. Let's get out in the open where we can see and be seen.'

With renewed caution we ascended the rocks, briefly turning a few over to find that scorpions were not rare, and several types of centipede also hid under the rocks by day, and one lizard shot off before we had time to examine it.

At the top of the ridge the land was flatter, and vegetation generally scarce since the water rapidly drained into the porous sandstone, only to emerge at the foot of the gorge in the pools we had seen earlier. A lot of loose weathered rock covered the surface, and clearly the animal life tended to hide in the crevices and under stones. There would be dew here most nights, and plenty of shelter for small animals. It would be fascinating to see the area at night, probably with a great deal more activity from the smaller animals than during the heat of the day that we were enduring with increasing discomfort.

Following the edge of the valley, we could look down on the vegetated areas by the pond we had passed on our way up, but now there were signs of life amongst the ferns and cycads. Small dicynodonts similar to the ones we had met were out foraging in the vegetation. Maybe they wait in the burrows until the sun warms the valley before they emerge, and we had passed by before the breakfast call. They were clearly happy to forage by day, possibly because they could bolt down a burrow if danger threatened. From our vantage point it was now obvious why they lived in the gorge; there was water, food and soft sediment for their burrows. Up here there was no water, little vegetation and virtually no soil deep enough for a dicynodont burrow.

Arriving at the junction between the Old Red Sandstone cliffs and the dunes, it was clear that there was an easy way back on to the dunes—one dune was conveniently banked against the cliff face—so with minimal clambering we could join the crest of the dune. However we were now more cautious. This was such an obvious route that it seemed unlikely, from our recent experience, that we would be the only ones to spot the opportunity.

A quick scan with the binoculars soon revealed that there were tracks in the sand, and this was indeed a regular route. There was also an obvious spot where animals could descend the face. As we neared this point, and even though we were walking on rock, there were signs of tracks and wear to the rocks, rather like the regular tracks used by sheep on a mountainside. Then we began to see the bones. A few were just scattered here and there, generally broken. Picking one up revealed clear bite marks, where sharp teeth had sunk through flesh to the bone. Next we came upon a leg, a short robust femur still attached to a tibia and with evidence of gnawing.

'Something carried this here from a kill,' Carol ventured. 'The animal was not killed here.'

'Maybe two animals then—a top predator, and a scavenger that drags bits of the carcass off to feed on at leisure or even to take home to the family.'

'Hmm, maybe.'

'We have to move on—but keep both eyes open.'

Cautiously we approached the point where the dunes lapped against the sandstone cliff line. We were in luck; there would be no problem getting down to the dunes. Clearly lots of other animals had done the same thing—tracks were everywhere. There was also a rotten flesh smell carried in the wind blowing up the slope towards us. Peering over the edge, a scene of some carnage met our eyes and nostrils. A ledge only five metres below was strewn with bones and fragments of carcasses, rib cages, limbs and parts of backbone; we were standing right above a lair. Before we could take in the scene, loud grunting and hissing sounds emanated from below, but we could see nothing.

'There must be an overhang, a cave, with carnivores.'

'And at least two are awake.'

We watched and wondered. Should we try and get a sight of this animal, or retreat? If we went down the 'path' to the dunes we could be spotted. The three 'how' questions arose: how many? how big? how hungry?

Retreating from the edge, we circled round to find a cliff-top vantage point from which we could see the ledge and its inhabitants. Having put a good fifty metres between us and the carnivores, through binoculars we could then see several animals lying in the shade beneath an overhang of rock; an overhang on which we had been standing. The animals were rather dog-like and a bit bigger than the two that had shot past us earlier; more labrador size but without the friendly wagging tail and good looks.

'They look smallish and quiet—not large carnivores, possibly scavengers. Let's go down.'

Cautiously we moved off down the slope from the cliff top to the dunes, and into view of the animals. Nothing happened for about twenty paces; then a sharp grunt rang out, almost like a dog bark. We looked but kept walking. There was now a line of at least eight of the beasts at the edge of the ledge. They were of several sizes. The largest stood out in front—his or maybe hers was the bark. The others appeared to be subordinate to the leader, and took up the call. This was a family pack, and they appeared to be aggressive, like a pack of wolves or hyenas. They had an enormous gape to the jaws, and repeatedly displayed the fact. We did not need binoculars to see that they had large sharp canines, exaggerated as in a sabre-toothed cat, and backed up with an impressive array of smaller jagged teeth.

'I think these are the predators: gorgonopsids by the look of them, and they are obviously quite robust and active. Those teeth could do serious damage, and I expect that they hunt as a pack. I hope they are not hungry.'

One of the beasts came closer, followed by two others; the jaw-gaping threats continued, but they did not seem inclined to attack.

'They seem confused; maybe our size and upright stance confuses them. We must not run and let them chase us. This is the face-off, so let's make ourselves big and fearsome,' I suggested.

What followed was a convincing primate threat response. A lot of jumping up and down, waving arms, and shouting and screaming. We paused, and the gorgonopsids fell silent. The leader looked round at the others, a clear sign of weakness. We advanced a few paces and started our display again. They had had enough, and backed off with backward glances. We held our ground for a while and watched them retreat towards their lair.

'Good, one to the bipedal mammals, I think.'

Soon we dropped behind a dune crest and were out of sight of the lair. We slid down the slip face to walk back roughly the way we had come. The dunes looked so different now; we were looking away from the sun, and they were fully lit, backed by the white of the playa lake. Sinuous crest lines snaked into the distance to east and west. There was no river channel through the dunes..In a big flood the weight of water would have to smash its way through, and erode a flood channel. Looking back up the dune we saw that we were still being watched. A lone animal stood on sentry duty at the top of the dune, but as we turned he gave a departing bark and disappeared back towards the lair.

After a trek of twenty minutes we began to wonder if we were off course, but we had been checking the compass, and it was a matter of some pride to find our footprints still visible on the harder surface between two dunes. Then we picked up familiar landmarks; the dried bones patch, the beetle burrows and the first trackway we had found.

Then there was something else. Our track was joined by a third track, and it was unlike any we had seen. This was not Pooh and Piglet being joined by a Wurzle, not unless a Wurzle is four-footed and can run rather well. The footprints were widely spaced along the track, and the trackway was quite narrow; thus this animal was not a splay-legged plodder, as were the herbivores, but carried the legs under the body and could probably run quite fast. The animal had been loping along the line of our track but following in the wrong direction. Was this just stupidity or reason? Could it not read our tracks and scent, or was it more concerned to find out where we had come from? It was clear that it was following the track, as its trackway would sometimes diverge from ours for a short distance, but then came back and followed our line accurately. The footprints were not large, considering the length of the stride, again pointing to an agile, athletic animal.

The Bus was not far away, for we were nearly through the dunes. If our Wurzle had followed the tracks to the dead end at the Bus we thought that it might retrace its journey. Rather than have a potential head-on meeting around the next dune, we deviated from the route and climbed the last of the dunes to find a viewpoint where we could see to the north, over the edge of the salt lake and hopefully a view of the Bus. We soon reached the dune crest and surveyed the vista of lakes and salt flats to the north. The Bus lay waiting for our return only a few hundred metres away beyond the edge of the dunes. Through binoculars our trackways were visible right back to the Bus, and the animal had followed them all the way. Goodness knows what this animal had made of our vehicle, but it had not stayed. A trackway set off along the lake shore, and there in the distance we could just see the animal, loping along through the heat haze. Maybe it was a large solitary gorgonopsid. We had just missed our mystery predator, only a distant tantalising glimpse with no chance of detailed observation. Fossil bones might be found one day to help solve the mystery of this animal; probably the top predator in the area in the late Permian.

Feeling rather disappointed at this inconclusive ending to our journey, but quite relieved that we had not met the animal at close quarters, we slid down the dune and walked over to the Bus. The animal tracks were all round the Bus, which had been well inspected, and there was a pungent smell. The strange object of the Bus had been sprayed by the animal, presumably as part of its territory, and I suppose that standing as it was in the open on the shoreline it was a rather obvious marking post.

'It's your turn to clean the Bus,' observed Carol.

Excursion 8

Dinosaur Dinner on Skye

TIME: Early Jurassic.
LOCATION: Shores of Loch Slapin, Isle of Skye.
OBJECTIVES: To visit a Jurassic seashore on Skye and watch life on land and in the shallow sea.
MODERN EVIDENCE: The ancient Jurassic shoreline features and fossils preserved in the rocks south of Camas Malag, near Torrin, Isle of Skye, and in the early Jurassic elsewhere in the Hebrides.

It could have been Western Australia. The sea lapped on a beach of white sand, and behind the beach rose low hills with rocky outcrops partly obscured by dark red gravel. The vegetation was fairly sparse by the sea, with cycads emerging like miniature palm trees above a general vegetation of ferns. Farther away a well-forested valley ran down to the sea, and a small stream had left a spread of brown sand that contrasted with the white sand fringing the rocky shore. Islands dotted the sea, some with pale jagged outlines of quartzites and others grey and more rounded in shape, probably made of limestone. In the distance, hills of Torridonian sandstone—part of the Scottish Jurassic island—rose above the narrow lowland strip that bordered the sea (Fig 8.1). The bedrock in the area where I was standing comprised the same rocks seen beneath the Jurassic around Broadford today: Torridonian Sandstone of Precambrian age, and Cambrian to Ordovician quartzites and limestones.

At Coral Bay in the far northwest corner of Western Australia, a visitor can walk on a rocky platform that is covered at high tide (Fig 8.2). At low-tide level the platform ends abruptly in a small submarine cliff on which corals grow, and a profusion of fish dart between corals and crevices in the rock face. Much farther offshore is a larger fringing coral reef with a steep dropoff into deep water. Large predatory fish—marlin, sailfish, Spanish mackerel and sharks—patrol the reef edge, and the sharks sometimes come close to shore, cruising the shallow water beyond the rocky beach platform. At the water's edge the beach sand which washes over the rocky platform contains grains derived from two sources. Some

Fig 8.1 Sketch of the early Jurassic coast at Camas Malag, near Torrin, Skye. Excursion 8 follows a route from the viewpoint to the beach and along the beach to the small delta. The hills of the Scottish mainland are made of Precambrian Torridonian Sandstone.

Fig 8.2 The coast near Coral Bay in Western Australia reveals evidence of the relative change in sea level. A narrow sandy beach fringes a rocky intertidal limestone platform rich in marine life. Corals grow at the edge of the platform. The low cliff is made of raised reef limestone and is being undercut by wave action and the boring activities of organisms.

are rolled fragments of shells from marine animals that inhabit the reef—corals, bivalves, gastropods, sea urchins and more. Other sand grains are made of quartz and are derived from sandstones that form the adjacent land.

Intermittently along the beach there is a low cliff, only three to four metres high but interesting because it is not made of sandstone but of limestone. This limestone contains corals and shells the same as those still living in the area at present, but here preserved as fossils in solid limestone rock. This limestone is the remains of an older reef, formed when the sea level was relatively higher. It could be that the land has been elevated, or that the sea level has dropped.

Back in Skye, in the early Jurassic, a similar situation existed. At the back of its Jurassic beach the bedrock was limestone, consisting of large blocks cemented together with sand containing marine shells. At some time the margins of the limestone blocks had been bored by marine bivalves that left distinctive, vase-shaped holes in the rock. Later, the holes filled with shelly sand and became cemented. These bivalves lived below low tide; thus these rocks must have been uplifted, or the sea level must have fallen. This could be confirmed because above the bored limestone blocks there was a metre of coarse shelly limestone containing shells rounded by wave action—a raised beach deposit. Since this deposit was now some three metres above high tide, it was clear that relative movement of sea and land had taken place, and as at Coral Bay the deposits formed at a higher sea level were now being eroded to leave a low cliff.

These features show that, when the Jurassic seas invaded the low-lying arid plains left at the end of the Triassic Period, the sea did not steadily rise and lap farther and farther on to the old Scottish land surface, but instead the sea level rose and fell in a series of pulses. The sea reached a higher level with each pulse, but between the highs the sea level fell and deposits of the previous high were partially eroded at the sea margin.

There was also a further contributory factor: the sedimentary basins of the west of Scotland continued to subside and fill with sediment; thus subsidence was creating space for the sediments to accumulate and be preserved. In Skye we had chanced upon the shoreline of the Jurassic sea at one point in this process. With time the sea transgressed farther on to the land, and the evidence that we can now see in rocks exposed at Camas Malag on the shores of Loch Slapin was buried for 190 million years. The area was uplifted by forces associated with rifting to the west of Scotland which eventually resulted in the opening of the North Atlantic, and locally uplift was caused by the intrusion of igneous rocks associated with the Skye volcano of early Tertiary time. Erosion since that time has quite fortuitously

Fig 8.3 View of the Cuillin Hills over Loch Slapin from the coast south of Camas Malag. In the foreground dark Lower Jurassic strata rest unconformably on steeply dipping, grey Cambro-Ordovician Durness limestone. In the background the mountain of Bla Bheinn (Blaven) is part of the eroded Skye volcano of Tertiary age.

exposed the ancient rocks of the Jurassic seashore on the present seashore of Skye at Camas Malag (Fig 8.3).

If we digress briefly south of the Scottish border, to the Mendip Hills of Somerset, a further example of this Jurassic sea-level rise can be seen. At the end of the Triassic Period the Carboniferous limestone we now see in the Mendips formed hills rising from an arid plain of dried muds with salt lakes. The slopes of the 'ancient Mendips' were scarred by valleys that drained the hills. However, these valleys seldom saw water, and were more like wadis in the desert, floored by gravel eroded from the limestone. The sea invaded this area in the late Triassic and sea-level rise continued into the Jurassic. During this process the 'ancient Mendips' became islands that were gradually drowned beneath the Jurassic sea. Thus in the Mendips, as in Skye, progressively younger layers of Jurassic rock came to rest on the slopes of the hills until the hills themselves eventually sank beneath the sea. An added interest in the Mendips is that the limestone was eroded to form deep fissures and caves, and as the sea level rose these were flooded and filled with sediment. The fissures contain fossils—some derived from the sea, but others representing animals that lived on the Mendip islands. There are bones of many small reptiles, and also very rare teeth of small furry mammals that were no bigger than mice.

In Skye it is likely that small reptiles and mammals also scurried over the Jurassic islands, but their remains are very rare as fossils. However, there is evidence from Skye slightly later in the Jurassic: a mammal tooth from near Elgol, and scattered dinosaur remains and footprints from Trotternish. These fragmentary fossils are in rocks originally deposited as sediments eroded from the nearby land, and along with some plant remains they give us a brief glimpse of life on the Scottish Jurassic island.

I had the Bus to myself for the day, and had chosen to visit one of my favourite Scottish localities, both for the rocks and the view. The Jurassic view was certainly different, but the sculpture of the distant mountains clearly indicated that they were made of Torridonian sandstones—the thicker beds forming distinct parallel features on the mountainsides. Standing near the Jurassic shoreline of a Scottish island there was plenty to catch the eye and excite the curiosity. The sea looked as attractive as any in a holiday brochure, and no hotels spoilt the view. All was totally deserted. The sea was inviting, clear water reflecting pale turquoise blue over shallow areas of white sand and darker blue where seaweed or rocks covered the seabed. To the north small islands rose from the sea, their coastlines a mix of rocky shores and white crescents of sandy beaches. Beyond, the slopes of the main Scottish island were clothed with lush vegetation at lower levels, grading upwards to a more scattered cover, and with rock exposures on the highest parts. Height was difficult to judge in the absence of familiar objects such as ships or buildings, but I guessed the land rose to more than 300 metres.

This was certainly a superb 'away from it all' location for a naturalist, a truly pristine natural environment. No litter, no boats, no vapour trails in the sky, but also no paths to follow. It was a case of picking a route through the rocks and vegetation to the shore. The bedrock here was limestone, solution weathered into bizzare shapes with sharp corners that could slice through skin. Many large blocks of limestone were detached from the bedrock and were unstable to walk on, a prime spot for cuts and sprained ankles. The crevices between the rocks also required care: ferns and small cycads crowded these sheltered areas, and it was easy to slip down the cracks. This was typical vegetated karst topography as seen today at Malham Tarn in Yorkshire or The Burren in Ireland. There were even small areas of karst at the present-day Camas Malag on Skye; however the Jurassic vegetation was totally different, and to an untrained botanical eye only the mosses and lichens looked familiar.

I left the Bus and scrambled over the rough limestone towards the shore. Nearing the top of the low cliff above the beach the vegetation reduced and the surface was easier to walk on. The cemented shelly deposits of a raised beach capped the cliff and were only sparsely vegetated, probably the result of salt spray and storms. The beach lay below—a glaring white strip of sand, interrupted by a few rocky areas, the whole contained between two low headlands a few hundred metres apart.

A sharp squeak heralded the arrival of the first animal, but where was it and what was it? Loud but small, a lizard with a mottled brown body and a bright red throat was adopting a threatening posture on a rock from which it had a good view of its patch. Another lizard answered from a metre away, then another, and suddenly I was in the midst of a miniature display frenzy. Turning to look around put all to flight, and they disappeared into crevices in the rocks. I must have been standing still long enough for them to feel I was part of the landscape, but movement produced instant evasion. They reminded me of lava lizards in the Galapagos, except that in those islands the lizards showed no fear of people, possibly because they had few predators other than the Galapagos hawk. This implied that there was maybe a ground-based predator here that cast a shadow worthy of fear. As there was no sign of anything large, it was time to press on.

Scrambling down to the beach was easy, and once on the sand evidence of animal activity was everywhere. Imprinted in the sand smoothed by the last tide were many trackways made by an animal using eight legs. Following the tracks was simple: they ran between landmarks on the beach—features such as pebbles or piles of seaweed. I traced them back to burrow openings above the high-tide mark, where piles of sand had been thrown out of the burrows. As with all wildlife watching, the best tactic was to sit still, be quiet and avoid making a silhouette against the sky. Eventually a cautious leg appeared from a burrow, then another, plus two eyes on stalks, then the whole crab. Others soon appeared and cautiously explored, looking for food on the sand and along the strandline, taking advantage of landmarks for cover, and maybe for navigation. As soon as I stood up panic erupted and they scuttled rapidly back to their burrows and disappeared from sight. These were Jurassic ghost crabs, related to those that are so common on beaches in the tropics today and which had diversified into many species suitable for different types of beaches and various kinds of food. In the fossil record the crab itself is rare, but the burrows they made are quite common and are closely comparable with modern examples.

With the crabs in retreat, a walk along the line of debris thrown up by the last tide provided evidence of what lived on and near the beach. The strandline was also a great source of food for a scavenger, and many animals would patrol the line of debris in a search for food. At first glance this strandline was little different from that on any modern tropical beach. There was seaweed, some fresh from the sea, some brittle and dry. There were pieces of wood, mainly branches and twigs, but stuck at the top of the beach were a few larger trees. Leaves of cycads and conifers were recognisable, as were some conifer cones. Shells and their broken fragments were everywhere: thick oysters, scallops and several shells of ovoid burrowing types of bivalves that must have lived in the sand down by the low-tide mark or beyond. It seemed that the basis of a good seafood menu was locally available here in the Jurassic, just as it is in Skye today. There were also gastropods, similar to our modern whelks and winkles; most of these would be living on rocky areas.

I started flipping over shells and seaweed with a stick, and soon found the normally expected creepy-crawlies. There were sand shrimps, and inevitably flies on rotting seaweed. I would be pleasantly surprised if there were no biting insects, but unhappy if they were as ferocious as the midges that can make a wet warm day a misery in Skye. Fantastic! Here in front of me lay the remains of a large ammonite shell about forty centimetres in diameter. It had been larger, but the body chamber in which the squid-like animal lived had been broken away. The chambered shell was encrusted with seaweed and worm tubes, and had clearly drifted in the sea for a long time after the animal died. The chambers of the shell contained gas, thus the shell floated, and could be transported a long way by ocean currents. The chambered shells of the modern pearly nautilus, a distant relative of the ammonites, are widely distributed from their living areas and cast up on beaches, and the same was probably true for ammonites.

There was movement farther along the strandline—I appeared to have the company of another beachcomber. A large lizard was attempting to eat something it had found on the strandline. From a distance I could see that it was tugging at something half buried in the sand. This appeared to be another reptile and to be long dead. I continued my walk towards the reptile and I was only about fifteen metres away when the animal clearly sensed my presence, looked straight at me and gave a serious head-nodding display, drawing itself up on the front legs to appear tall and frightening. I stood still and the display died down, but any movement set it off again. Advancing towards the lizard increased the vigour of the threat until suddenly it stopped, and the animal turned and ran, fast, straight for the cliff. It went up the low cliff and straight into a hole near the top. It ran like a goanna from Australia, but still on all fours, rather that the technique adopted by the 'racehorse goanna' that rises on to its rear legs and becomes bipedal when fleeing.

This Jurassic animal used this hole frequently: there was a regular trail to the entrance, and piles of dirt that had been excavated with strong claws. On the beach it had been tearing at a rather smelly carcass, and judging from the mêlée of footprints it was not the only visitor. But why did it run? Maybe it was poor at judging distance, and as I came closer it realised that I was tall, and to it tall beasts spelt danger. So was there something larger here that patrolled this beach, and was partial to a lizard nearly two metres long? Something about my length, come to think of it, but I carry more meat and lack the tail. I continued along the beach keeping a roving eye for any sign or sound of large animals.

In some places sand had been blown up the beach to form small sand dunes in the lee of the cliff, and delicate ripples of blown sand adorned the surface. There was evidence of animals in the form of trackways: most were small and probably made by lizards, while others were more like the tracks of the large scavenging lizard. There was also a larger area of disturbed sand, a depression a metre across, with large scuff marks, and evidence of digging. The giveaway was the broad trail that led back down the beach as far as the last high-tide line. There was a shallow, wide central groove, with regular impressions of paddles along the sides, and the sand was pushed into ridges where the animal had hauled itself over the sand. This was a turtle trail, and the only reason a turtle had to leave the sea was to dig a nest for egg laying. The nest was clearly fresh, made in the last day, and had not been raided by an egg eater as yet, but I would have been surprised if our large burrowing lizard was not partial to turtle eggs.

A few metres farther along the beach the evidence of predation of turtle nests was obvious. The sand had been trampled, so much so that individual footprints were not clear, but dried fragments of leathery white turtle eggs stained by dried yellow yolk told the sad story. At first it was surprising to realise that turtles had survived to the present day in the face of such predation pressure, but clearly enough eggs hatch and enough hatchlings reach the sea and grow to breeding age to perpetuate the ancient turtle line. This was a case of a seemingly vulnerable animal group that had survived and evolved for more than 200 million years. However, it was still sad to see the carcasses of tiny defenceless hatchlings that had been picked clean, even by the predators and scavengers of a Jurassic beach.

This looked a good place for a snorkel swim: the beach was sandy and open with some underwater rocky areas showing as dark reflections on the sea surface. Leaving my kit on the beach I ventured into the shallows. The sea was warm, and small waves lapped against my legs. This could have been any warm seaside holiday location, and it was difficult to accept that I could have been in serious danger. I am used to the beaches of the UK and the Mediterranean where there is little to worry about, but in other places more danger lurks. Apart from the spectacular, such as large sharks, there could have been many animals here that would have posed a danger to a mammal with a soft thin skin not suited to a life in water. The greatest menace lay in the animals I might step on or bump into. They were not likely to be interested in me, but I might unwittingly annoy them.

Animals with poison are the greatest hazard. Jellyfish, such as the modern box jellies of northern Australia, can deliver enough neurotoxin to stop a man breathing. It would be dangerous even to touch the long trailing tentacles. Some corals can also deliver a nasty sting from contact. Then there are fish with poison spines that half bury themselves in the sand, and can be stepped on with painful results. There are also types of octopus and gastropod that have bites or stings that can be fatal. Australia has more than its fair share of venomous sea life, but deaths are very rare. Every year far more Australians die by falling off rocks whilst fishing than are killed by all the dangerous animals of that great island put together, including venomous land snakes, spiders and crocodiles.

Comforted with these thoughts, and confident that at least some of these nasties were not about in the Jurassic, I launched myself gently from the shore. Swimming was safer than walking in water with bare feet, and through the mask I could only see clear water—no drifting jellyfish to avoid for now. The sea floor beneath me was covered in sand ripples, and as I hung in the water rocked by the waves, the sand below me was washed back and forth in flurries across the ripple crests. There was little life to see on the bare sand—any animals would be safely buried out of sight. Farther from shore the water deepened and the bottom took on a rougher texture; the ripples were gone, and the sand was clearly disturbed by animal activity. There were surface trails and open burrows, and a few fish were searching for food, digging in the sand and disturbing small shrimps from their burrows (Fig 8.4). The spire of a gastropod was just visible at the end of a trail where it ploughed its way through the sand, and a few empty bivalve shells littered the surface. Presumably the shells were dug from their burrows, the animal eaten and the shell discarded.

Fig 8.4 Sketch of submarine early Jurassic diorama with turtles, fish, belemnites and ammonites. The sea floor is colonised by algae, bivalves and crinoids. An *Ichthyosaurus* skeleton is partly buried and provides a foothold for algae and worm tubes. Shrimps scavenge the sea floor from their burrows.

Farther from shore the sediments appeared more muddy, and the surface was crowded with oysters. They had one strongly coiled shell and a smaller one that formed a lid. Most lay with the lid uppermost, and some revealed a crescent of yellow around the edge, showing that the shell was open and the animal was pumping water through its mantle cavity and extracting food and oxygen from the water. These were *Gryphaea*, a very common Jurassic fossil at Broadford and south of Camas Malag (Fig 8.5). I could see a small crinoid attached to one oyster, and a few other bivalves, resembling mussels, were also attached to some shells. I was now a good 100 metres from the shore, and details of the sea floor were becoming vague due to the depth, so I turned to swim back and drifted parallel to the shore.

A curious shape on the sea floor caught my attention. It was partially masked by seaweed growth and encrusted with a few worm tubes, but I could clearly make out a line of vertebrae and an elongate skull with the empty orbit of the eye blankly staring back at me. It was an *Ichthyosaurus* skeleton, something I had longed to find as a kid on holiday near Lyme Regis, but never saw more than the odd bone. Here was the real thing in the process of burial. This marine reptile ate fish and also squid, catching them with its long toothy beak. It was very similar to the modern dolphin in body plan, showing how a reptile and a mammal can evolve to similar shapes from very different ancestors.

Fig 8.5 **A:** Early Jurassic muddy sandstones of the Broadford Beds crowded with the oyster *Gryphaea arcuata*. Shore at Waterloo, Broadford. **B:** Specimen of *Gryphaea* extracted from the rock.

Leaving the *Ichthyosaurus* behind I drifted over the edge of a rocky area, with tufts of green and brown seaweed. The sediment around the rocks was coarse and pebbly, and there lying on the surface were scallops, and of a good edible size. With strong ribs and shells in brown to white colours they could almost have passed for the modern version—scallops having changed little since the Jurassic, and were probably just as tasty. On the rocky area there were sea urchins grazing the algae, and fish biting morsels from the rock surfaces. I had no idea what the fish were, but the bright colour patterns they displayed would not have been out of place in modern seas.

A large dinner-plate-sized fish appeared to my left: it was dark brown above and silver below, with blackish vertical bars on the flanks. The pattern blended beautifully with the seaweed strands growing from the rocks. The tail was quite small, and the dorsal and anal fins formed long fringes above and below the rear end of the disc-shaped body. The shape was similar to many modern reef fishes such as angel fish. The mouth was small for the size of the fish, and strong teeth protruded from the jaws. The fish repeatedly picked at the surface of the rocks, detaching small molluscs and crunching them in its jaws. It could pack quite a bite—I could hear the crunching underwater, just as a modern parrot fish can be heard crunching material from rocks and coral. This fish looked like *Dapedium*: it certainly had the heavy rhombic scales that preserve so well in fossil specimens. I was drifting nearly above the fish, but as I moved my arms to swim the fish darted off and disappeared behind a rock. It was hiding from me and clearly did not stay in the open when large objects swum past.

At last! A big ammonite came into view. This was one of the typical early Jurassic forms that can be seen in rocks of this age at Lyme Regis in Dorset, and also around Bristol, as well as in Skye (Fig 8.6). It is about fifty centimetres in diameter, and the heavily ribbed shell was coloured red and green. The animal was carefully searching the sea floor, probing in burrows with its tentacles in the hope of a shrimp meal. This looked like the living version of the dead shell I had found earlier on the beach. Ammonite shells are similar to those of nautilus in

Fig 8.6 Specimen of a large early Jurassic ammonite (cf. *Coroniceras*). From Broadford, Isle of Skye.

that both were occupied by cephalopod molluscs, and had a chambered shell with an open body chamber in which the animal lived. The chambers in the shell were connected by a tube called the siphuncle through which the animal could control its buoyancy by varying the quantity of gas or liquid in the chambers.

As I started swimming again my sudden movement revealed a small shoal of squid, previously unseen on account of their excellent camouflage. As they cruised away along a gully in the rocks their colour changed to match the background. In a flash there was commotion in the water, and the squid fled leaving clouds of black ink in the water. Gradually the ink drifted away, revealing a fish struggling to subdue one of the squid. Come to think of it, the squid were probably belemnites. The bullet-shaped belemnite fossil is part of the internal structure of a squid-like animal (Fig 8.7) and is partly equivalent to the modern cuttlefish bone so common on shores today. The fish had the belemnite gripped near the tail, but the belemnite had wrapped its tentacles around the body of the fish, making it impossible for the fish to manipulate the belemnite and swallow it. Suddenly the fish squirmed violently and opened its jaws; the belemnite had managed to bite the fish with its horny beak. Now it seemed that the fish wanted to give up, and it tried to regain the shelter of the gully from which it had sprung the ambush, but the belemnite, despite its wounds, refused to let go and was dragged along, still clamped firmly to the fish. The fish squirmed again; maybe the belemnite was

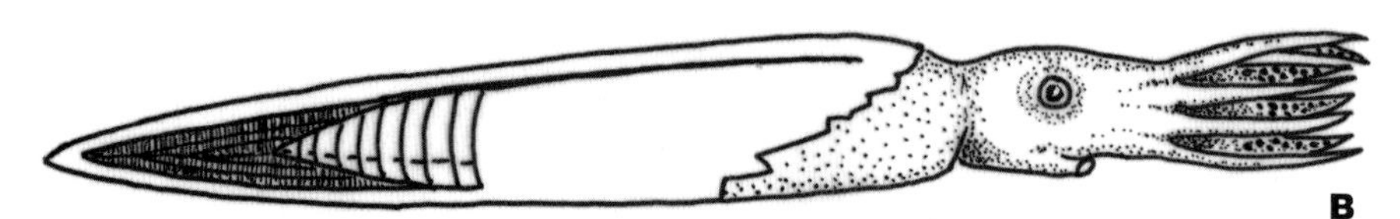

Fig 8.7 A: Belemnite fossil. **B:** Reconstruction of a belemnite. The fossil guard has part of the chambered shell inserted in the end, and is encrusted with worm tubes that grew on the belemnite as it lay on the sea floor. Belemnite fossil 11 cm long. Specimen from the late Jurassic of Staffin, Skye.

taking revenge and was slicing away at the flank of the fish. The fish was weakening, its tail thrusts were now spasmodic, and its mouth and gill covers were opening wide as if in a spasm. At last the belemnite released and swum gingerly away to hide in the rocks. Surely it must be badly bitten, but maybe it could recover given time and peace; cephalopods have a great ability to mend wounds.

The fish appeared to be in serious trouble. It had a small wound, leaking blood, and had many round sucker marks on its body, and it could not swim. It could not maintain an upright position in the water, and swam in random bursts, like an automaton out of control. Poison! That's it, the belemnite had a venomous bite, and the fish was partly paralysed. That explained why the belemnite did not let go of the fish.

Suddenly my heart leapt; a large black shape shot past from behind me; panic! I shipped water down the snorkel and spluttered for air at the surface. Below me the shape circled; it was a familiar shape—a shark shape. I should have predicted this: a blood-letting fight putting scent into the water, and a shark had swum up the trail to the source. This might only be the first one. I wished I had a wetsuit. Where was the shore? Answer—about eighty metres away. I needed to retreat gently, without splashing and panic. Looking under water again I was reassured. The shark had the fish across its jaws and was shaking it back and forth. Actually it was not that big a shark, probably only about two metres long, and clearly a fish eater, still big enough for an underwater surprise, but no real cause to panic.

Nothing followed me towards the shore. I poked my head up again to check the direction: good, only about forty metres to go—but another head popped up and stared at me from about ten metres away. My heart leapt again, but only for a moment; this was a turtle. Now I could see her under water—it was clearly a female. She swam lazily past gliding on outstretched front flippers and with no sign of concern or interest in myself. She was probably waiting for tide and darkness before hauling herself up the beach to lay her eggs. Once I was in the warmer water of the shallows I cruised back over the sandy ripples to the shore.

Then I sat down, dried off and reflected. I was feeling more relaxed in air than in the water. It was time to go farther along the beach and investigate the mouth of the burn that emptied on to the beach from a green valley. From a distance it was clear that the burn had brought sand to the beach and formed a small delta protruding into the sea. At most it looked to be 200 metres wide, but the vegetation came close to the water's edge, leaving a marshy area between land and sea. Approaching this little delta, the sand changed from a white shelly sand to yellow and brown, reflecting the iron-stained quartz grains brought down by the stream from the sandstones forming the hills farther inland. There was also more mud, sufficient to form small mudflats in areas sheltered from waves and river currents. Around the burn and up the valley the vegetation was a jungle of conifers, cycads and ferns, and from a distance the jungle looked impenetrable.

Sitting on a rock and scanning the area I could not see any animal activity. It was early afternoon, and my hope was that any large animals were likely to be resting. There was some water flowing out of the burn, but it looked as though it would not be a barrier to progress along the beach. A buzzing near my feet interrupted my survey, and I jumped away from the rock. A large clumsy insect was struggling to emerge from the sand disturbed by my feet. I had probably destroyed its resting place, and it was not pleased. I was just examining it from a safe distance for any sign of stinging or biting potential when the unfortunate creature had its life terminated. A small furry animal the size of a hamster, but thinner and with a long head and toothy mouth, had grabbed the insect and crunched its way through the unfortunate's thorax. Suddenly it darted back under the rock whence it came with its prize. Apologies, insect—but at least it had contributed to the rise of the mammals, so I had to feel grateful.

Approaching the burn the shore became muddier, and I stepped up on to the firmer ground of the marsh above high-tide mark. There was a mat of plants, and I could not put names to any of them; all that was obvious was that there were no grasses, and none of the flowering plants typical of the present day. There were plenty of small, spreading, fern-like plants that had joined up to make a patchwork of green and grey, springy underfoot and aromatic. Sometimes the tide flooded this area; there were strandlines with seaweed, driftwood and a few shells. My route was being funnelled by the sea to my right and forest to the left, but hey, I was on a path. It was trampled flat, about a metre wide and forming a slight groove in the marsh top. Others had passed this way, and did so on a regular basis. Looking back, the path appeared to have emerged from a low tunnel in the undergrowth; that was a relief, as it could not be the route of a tall beast. After a while the path headed to the shore, and I followed out of curiosity. The path dropped down to the beach through a groove in the marsh edge and was lost on the beach. Any tracks must have been obliterated by the last tide.

But what was this? The hairs on the back of my neck rose slightly. I was looking at a large, three-toed footprint in the mud at my feet. It was impressed deeply into the muddy sand that had squeezed between the toe prints; there were marks in the mud where the animal had lifted its toes. I did not need any experience with trace fossils to be able to interpret the presence of a long sharp claw on the end of each toe. This place was not so attractive for a stroll any more. The print was fresh, since the last tide, and the maker could not be far away. There was the next print, and there another. Its stride was almost twice as long as mine: it walked on two legs, and it was probably taller than me. I was trapped between the seemingly impenetrable jungle on one side and by the sea on the other. I was clearly vulnerable. This was far more nerve-tingling than the time in Montana when lazily walking along the marshy shore of Hebgen Lake, casting a fly for the local trout, I had come across large fresh bear prints that came out of the forest to the water's edge, and then returned to the forest. The bear had come down to the water for a drink. I had whistled nervously for a while, and kept more of an eye on the forest than the rising trout.

But this was different. A solitary, large, patrolling, bipedal reptile with large claws was not something I needed to meet. Was it wise to follow these tracks? The mouth of the burn was just round the next corner, screened by vegetation; a fallen tree had toppled on to the beach and blocked the view. I would go that far and then turn back. After all, the really big flesh eaters like *Tyrannosaurus* did not appear until the Cretaceous. However, the first dinosaur to be given a name, *Megalosaurus*, was a large carnivore from the Jurassic, but the oldest one known from Britain was still a good 20 million years younger than this time in the Jurassic. But it must have had ancestors—had the fossils been found? Anyway, I had seen the evidence; the beast was not a five tonne monster, but even something the height of an ostrich and sporting a toothy grin was enough to set alarm bells ringing. Carnivorous bipedal dinosaurs 3 m long were around in late Triassic times in Argentina, so there had been time for a serious predator to evolve.

I was now walking with care, trying not to make any noise; soon I would be able to see around the corner, past the fallen tree. At least the wind was in my face, so if there was something around the corner it should not smell my approach. The fallen tree was worn smooth from the action of waves and sand, and sprawled over the beach still fixed to the earth by twisted roots at the top of the beach. Branches pointed skywards, still adorned with withered leaves. Peering through the branches, I could see the river mouth with small sand bars around which the channel divided; farther down the beach the river water spread out and was lost from sight amongst the pebbles. Above the beach the gently sloping land was covered with ferns and cycads, and a few larger conifers like the one I was using as an observation post. I could not see much because the ferns were more than a metre high, and I had a low vantage point on the beach. However, I could hear animal activity, and I fancied I could smell animal as well. There was an intermittent grunting coming from the ferns. Distance was difficult to judge. Was it a small animal close by or something larger and farther away?

Needing a better view, I grabbed a branch of the tree to haul myself higher. The tree moved under my weight, and screeching and clattering broke out above my head. Instinctively I let go and half jumped, half fell back on to the beach, as dead twigs showered down on me. Above me there were three flying shapes, flapping away on bat-like wings. They were about the size of a jackdaw, with scrawny necks, and a head with large eyes and a mouth with visible teeth. A long thin tail served to aid balance as they flew above, chattering rather as starlings do. They were not birds but flying reptiles, maybe *Dimorphodon* or a relative. Calming down from the shock I realised that any animal within sight or sound of this event would be wondering what had disturbed the sleeping *Dimorphodon*, which deciding I was not a threat, came back to the tree. Hanging on with both hind feet and finger claws, they again took up positions on branches, all in head-down pose, presumably to aid takeoff. They were now alert and fixed me with a stare from large dark eyes surrounded by red rings, and protruding from leathery eye sockets. They looked distinctly sinister and could have acted as extras in any horror movie. It was interesting that the inverted hanging pose adopted by modern bats and flying foxes was already perfected by reptiles in the Jurassic. Further thoughts on these beasts were interrupted by more grunting from the ferns. Quickly back on my feet, I peered over the tree trunk again.

There was no need for a higher vantage point now: a neck and head protruded well above the ferns, and the head was looking my way. Time to stay still. The head was bigger than mine, with a mouth full of sharp teeth, and the blood over its face and trickling down its neck did not give cause for comfort. It moved its head from side to side slowly as if trying to make out a 3D image. It took a couple of steps towards me—it must be a good three metres high when standing upright. Now I could see the front legs, not very long, but clearly with functional, and literally bloody claws.

Then it lowered its head, opened its mouth wide and gave out a guttural hiss, which seemed like a threat or a challenge. Still I was not convinced it could see me. In response to a high-pitched series of squeaks it turned its head, immediately roared, turned and charged away from me. A couple of small bipedal dinosaurs no bigger than turkeys shot out of the ferns and on to the beach, but the large beast did not follow. It was protecting a kill. In the circumstances I did not think I was going to try and get close enough to see the kill; I did not want to become the second course. I then knew what the smell was—more the kill than the dinosaur.

I also realised that there was potential danger behind me. If I could smell the kill, reptiles would probably be able to scent it a thousand times more clearly. Where would they come from? Behind me? As I sat by the fallen tree, one could even now be following my tracks along the shore. What of the animal pathway I saw earlier? I felt trapped already.

It was uncomfortable crouched behind the fallen tree, and since the beast had its head down I shifted position gingerly to get the circulation going in my leg. As my hand touched the log I felt movement and instinctively drew my hand back. A large insect spread its wings and buzzed into the air. It did not get far. The *Dimorphodon* immediately started an aerial attack, launching themselves into the air virtually simultaneously, like gulls after a piece of bread thrown from a seaside pier. The insect was a heavy flier and was soon grabbed with a loud snap of *Dimorphodon* jaws. As with gulls, a chase now ensued, the others trying to rob the catcher of its prize. The agility was surprisingly good, but not up to the flying skills of a bird. The insect having been consumed, the hunters returned to the tree above me. This was clearly a favourite hunting perch, and I had flushed a tasty morsel.

I could hear distant bellowing, a new sound, and coming from behind me. But I could spot nothing close at hand. With regular glances in the direction of the kill I scanned the distant scene. Movement! A long way off, on the distant shore of the mainland, a group of sauropods with long necks and tails were wandering along a beach. They were rather decorative with strong vertical stripes of yellow and green on the sides of the body and a reticulated pattern in the same colours on the neck, back and tail. They were not the giants of the late Mesozoic, more like six to ten metres long and looking at a distance like *Plateosaurus* of the late Trias. I could see about a dozen of them, of different sizes, so maybe they were a family group. They plodded along peacefully—possibly the beaches were major highways for animals in this area.

My eye then caught another and nearer movement, and there it was—another carnosaur, possibly bigger than the first. It was walking purposefully along the beach towards me, 150 metres and closing. It sniffed the air regularly, hopefully scenting the other carnosaur's kill, and not me. I could only sit tight and hope; to move now would probably be fatal. At 100 metres, I could hear the crunch of the sand under its feet, so I crept right under the fallen trunk. At fifty metres, it scented the air again, and the crunch of its footfalls stopped; it had gone up on to the marsh top. I could not see it any more but I could hear the breathing and snuffling. A roar from the kill site produced a reply from the new arrival. I looked up and saw powerful legs and clawed feet crash past. All hell broke loose as the two beasts met. Peering out, I saw them rearing up in front of each other and striking with jaws gaping. The newcomer was larger, and noisier, and to my surprise the owner of the kill backed down and retreated. Maybe it was size domination, or possibly the owner of the kill had fed enough—its belly looked seriously distended. The new arrival took over and started tearing at the carcass. It lifted part of the carcass in its jaws and shook the kill strongly, teeth cutting deep into flesh. I noticed the green-and-yellow, reticulated pattern of the sauropods that were wandering on the distant beach; it seemed that the *Plateosaurus*-like sauropod was a prey item. The new arrival was engrossed in the carcass, and the other one had retreated out of sight into the vegetation. It was time to make a break for safety and go back to the Bus.

Cautiously I extracted myself from under the tree and crept away. The beast did not look up, and the *Dimorphodon* remained silent. Once out of sight of the kill, and feeling some relief, I was suddenly distracted by noises from the vegetation. Something was coming this way, and quickly. I was near the animal pathway that I saw earlier leading to a low hole in the vegetation. Something was going to come out of that hole. I knew it!

It appeared at speed, and was instantly recognisable. Crocodile! Having seen large saltwater crocs in Queensland rush to the water out of mangrove forest and slide down the river banks into the water, I knew what to expect, and also that it was not sensible to be in the firing line. However, fear rapidly subsided when I noticed that the crocodile was only about 1.5 metres long, and had narrow jaws suitable for catching fish. It slid down the bank and shot into the water. Something must have frightened it from its resting place, but there did not seem to be anything following it. I too made my way down to the shore and walked back along the water's edge. I reckoned that the most danger lay on my landward side, and should anything appear I would take to the water for temporary refuge.

Thankfully my walk along the shore passed without incident, and I was soon safely back at the Bus. This quiet walk had been rather more stressful than anticipated, and it seemed there was the potential for more exciting dinosaur finds on Skye.

Excursion 9

Amorous Ammonites

TIME: Mid Jurassic.

LOCATION: Bearreraig Bay, Trotternish, Isle of Skye.

OBJECTIVES: To observe ammonites in their natural habitat.

THE MODERN EVIDENCE: The rich ammonite fauna and other fossils preserved in the sandstones at Bearreraig Bay, Trotternish, Isle of Skye.

The Jurassic and Cretaceous seas that surrounded the 'Scottish island' were warm and shallow, and home to a rich fauna. The climate fluctuated, but was generally warm and temperate, something akin to the northern shores of the present Mediterranean. Scotland marked the approximate northern limit of coral growth at the time, and deposits were sand and mud, now mostly sandstones and shales. Farther south the sea was warmer and more limestone was deposited, such as the rich honey-brown oolites of the Cotswolds from which villages such as Broadway are built, and the sterner white to grey Purbeck limestone, so prominent in the buildings in London.

The elements of the marine fauna inhabiting those pleasant shallow seas that were most suited to preservation as fossils comprised: the molluscs, with a variety of bivalves, such as oysters and scallops; the gastropods—marine snails, of various shapes and sizes; and the cephalopods. The majority of cephalopod fossils of the Jurassic fall into one of three groups: ammonites, belemnites and nautiloids. Of these groups the ammonites and belemnites are extinct, but nautiloids hang on with only four known species left in existence, still with shells incredibly similar to their ancestors of 200 million years ago. Both the external form of the coiled shell and the internal division of the shell into chambers connected by a tube are the same today as they were in the Jurassic (Fig 9.1). This is a classic example of a conservative lineage. However, this lack of any evolutionary innovation has put the nautilus on the very edge of extinction—a process being accelerated by man due to the attractive shape and colour of the shells. It is a great shame that the shells appear in sad rows in seaside tourist shops, on sale for paltry sums.

Fig 9.1 Modern and fossil nautilus shells. **A:** Modern shell. **B:** Shell sectioned to show the chambers and perforations in the chamber walls that carried the siphuncle, a tube connecting the chambers (body chamber missing). **C:** Broken Jurassic fossil nautilus revealing the chamber walls and siphuncle encrusted with calcite that does not completely fill the chambers. **D:** Fossil nautilus from the mid Jurassic.

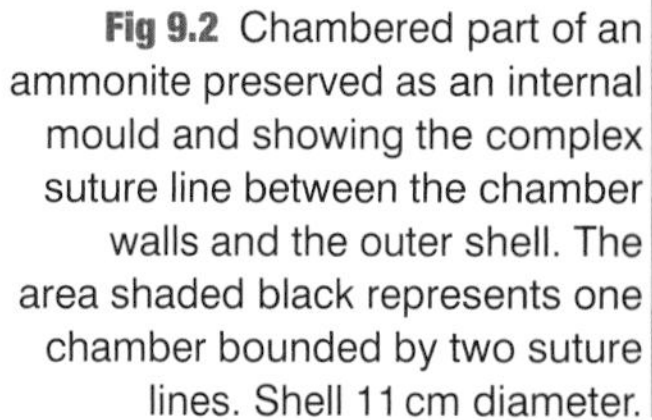

Fig 9.2 Chambered part of an ammonite preserved as an internal mould and showing the complex suture line between the chamber walls and the outer shell. The area shaded black represents one chamber bounded by two suture lines. Shell 11 cm diameter.

The animals are being trapped in their few remaining strongholds, such as the Philippines, where there is little regulation to protect a shell for which there is a ready market. Before the advent of diving the living areas of nautilus shells were unknown, and most shells had been found on beaches. Many nautilus shells float after the death of the animal and drift on currents to be cast up on far-flung Pacific and Indian ocean shorelines. In earlier times the shells were greatly prized, and intricately carved to reveal the mother of pearl shell layer, and expensively mounted in silver as high-class ornaments. That at least was a fate of greater respect than having their apertures filed down and then sold for a few dollars in any currency as a seaside souvenir.

Zoologically and geologically the nautilus is a fascinating animal, providing us with the proverbial 'living fossil' that can be used to interpret the zoology and life habits of the fossil examples. Further than that, they are the only cephalopods with external shells still living in our seas and oceans, and thus they are also our best link with the fossil ammonites.

What the nautiloids lacked in evolutionary creativity was possessed in abundance by the ammonites. Their chambered shells display a bewildering variety—fat or thin, smooth or ribbed, knobs, spines and spiral grooves. The shell ornament, as well as the mathematics of shell coiling, provide rich resources for the study of morphological detail. A major difference between nautilus and ammonite shells is that ammonite shells have a complex suture line. This is the line formed by the junction of the chamber wall with the inside of the shell, and it can be seen in fossils where the outer shell is missing (Fig 9.2). In nautilus this line is a simple curve.

Ammonites evolved rapidly in the Jurassic and Cretaceous periods, and thus they are used as zone fossils, enabling the relative age of the rocks to be determined with great accuracy. In parts of the Jurassic, ammonite 'zones' can be distinguished that lasted for less than 250,000 years. Thus the evolutionary energy of the ammonites was great, but this did not help them to long-term survival. Despite the advent of bizarre forms and the continuation of conservative shell types, they all died out at the end of the Cretaceous. So the 'stick in the mud' nautilus has had the last laugh, a laugh lasting for 65 million years, yet it has few close relatives with whom to celebrate.

Belemnites, one of the most common fossils of the Jurassic, are parts of the internal skeleton of a squid-like animal, broadly equivalent to the tiny horny knob on the thin end of a cuttlefish bone. Whilst belemnites do have their academic followers (maybe they also collect different types of ammunition for variety) they are a rare breed. On the other hand the beauty and diversity of ammonites have attracted palaeontologists for a very long time—Plinius the elder who died in AD79 called them *ammonis cornua* (meaning 'Ammon's horns') after the Egyptian ram-headed god Ammon with his spiral horns.

A fascination concerning both ammonites and belemnites is that they can be extremely abundant. Some beds in the Jurassic and Cretaceous are literally crammed with ammonites, and other beds are so full of belemnites that they have been called 'belemnite battlegrounds' from their resemblance to so many discarded bullets. In a strange quirk of fate, on a patch of beach near Brora on the east coast of northern Scotland, belemnites can be found washed from the Upper Jurassic Brora shales. Mixed with them is the occasional bullet—presumably a relic of wartime exercises in the area and the nearby abandoned shooting range.

Concentrations of ammonites and belemnites in certain beds could be due to a variety of factors. They may have accumulated during a period of slower deposition because of reduced sediment supply; and they might have been swept together by current action in the shallow seas. Another possibility is that, particularly in sandy sediments, they were only fossilised in favourable conditions, and the fossil beds are of preservational origin—there having been sufficient carbonate to cement the bed and prevent solution of the calcareous shells during burial. A more speculative explanation is that such accumulations represent the mass death of the animals following spawning. Modern squid are known to gather in specific areas to spawn in vast numbers, and die following spawning. If the cephalopods of the Jurassic behaved in the same way, our belemnite battlegrounds may be more the result of battles of the sexual kind, with the struggle directed to the passing on of the genetic code.

Some of the best examples of ammonite and belemnite concentrations in Scotland occur in the foot of the dramatic cliffs at Bearreraig Bay on the Trotternish Peninsula of the Isle of Skye (Fig 9.3). The road from Portree to Bearreraig Bay winds north below the great escarpment of lava flows poured out in Tertiary times some 55 million years ago. Looking back south from this road, the Cuillin Hills represent the roots of a great volcanic edifice of Tertiary age, now deeply eroded. Our road roughly follows the unconformity between Tertiary lavas and underlying sandstones and shales of mid-Jurassic age. There is a complication in that the Jurassic is intruded by a Tertiary igneous sill of dolerite wih strong columnar jointing. This sill is resistant to weathering and forms the top of the sea cliff for much of the eastern coast of Trotternish. Before reaching Bearreraig the classic picture-postcard viewpoint of Loch Fada is passed and the Old Man of Storr—a

Fig 9.3 View looking north at Bearreraig Bay, Trotternish, Skye. The beds containing the ammonite *Ludwigia* occur at the base of the cliff at the left, and the main cliff across the bay consists of mid Jurassic sandstones capped by a thick Tertiary dolerite sill with columnar jointing.

column of Tertiary lava preserved in the landslip below the lava scarp on the skyline. At the far end of Loch Fada a narrow road leads right to the Storr Lochs dam and hydroelectric station, from where the sea cliff of Jurassic sandstones capped by the dolerite sill can be seen. As long ago as 1819 the Scottish geologist John MacCulloch in his *Description of the Western Isles* drew sketches of the sills intruding the Jurassic rocks of this coast.

A stepped path leads down to the beach, and for the object of our interest it is necessary to turn south at the bottom and clamber around the point and on to the beach. The rocks are slippery with green algae, and the point becomes rapidly impassable on a rising tide. Many a visitor scouring the beach for fossils has received a good soaking after failing to watch the tide. The ledges at the point mark the junction between grey shaly sandstones below and brown sandstone with rounded calcareous concretions in the cliff above.

This is a protected site (Site of Special Scientific Interest), so hammering of the cliff and ledges is not permitted—and since the best fossils are usually found with less effort in loose boulders on the beach, there is no reason to attack the outcrop.

In the ledges there are fossil-rich patches that contain an abundance of ammonites, mostly of the genus *Ludwigia* (Fig 9.4). The type specimen of the

Fig 9.4 Specimen from the Murchisonae Bed at Bearreraig Bay containing abundant ammonites, a bivalve (top right) and a piece of the chambered part of a belemnite (lower right; also see Fig 8.7). The ammonites have the chambered part of the shell filled with pale brown calcite, and the body chamber filled with darker sediment. Specimen 17 cm wide.

zonal Jurassic ammonite *Ludwigia murchisonae* was collected here by none other than Lady Murchison. She was the wife of the redoubtable Sir Roderick Impey Murchison, author of *The Silurian System* (1839) and, with others, produced *The Geology of Russia in Europe and the Ural Mountains* in 1845. This work contributed enormously to the development of European geology at a time when travel was difficult and time-consuming. However, Murchison was a person of great political power in the developing science of geology, and his (incorrect) theories on the development of the Scottish Highlands hampered progress for many years. As a geologist in the Geological Survey it would be more than a job was worth to disagree with Murchison's opinions; he was director of the Geological Survey.

Ludwigia murchisonae is a macroconch (large shell) form and is the size of a small dinner plate; its outer whorl is smooth and the shell margin at the aperture forms a simple curve. Also present are smaller microconch (small shell) forms that are about five centimetres in diameter, have numerous ribs and an aperture with a projection or 'lappet' at each side (Fig 9.5). These forms are considered to be

Fig 9.5 Specimens of the ammonite *Ludwigia*. The larger (macroconch) specimen is a typical *Ludwigia murchisonae* and is interpreted as a female, and the smaller, ribbed specimen with lappets at the aperture is a microconch and is interpreted as a male. Specimens from Bearreraig Bay, Trotternish, Skye.

sexual dimorphs, with the female being large and plain, and the male small and pretty. This is in tune with dimorphism in modern cephalopods where the male is frequently smaller than the female. Indeed, in one type of squid the male is reduced to little more than a sac of sperm, and having found a mate, spends his life fused to the head of the female—abandoning any chance of having a night out with the boys, or straying from the straight and narrow.

Along with the abundant ammonites in the fossiliferous patches of rock there are less common belemnites, bivalves and gastropods that lived in the sea, and pieces of fossil driftwood from the Jurassic forests of the Scottish mainland. The fossils in these rocks occur in patches that represent areas of calcite cementation of the sand, and the calcite has helped prevent solution of the shells by fluids passing through the pore space in the sandstones. At one time the bed was full of fossils, but now only the fossiliferous patches remain.

An obvious feature of the fossiliferous patches is that the ammonites are at all sorts of angles, not lying flat as might be expected if the shells came to rest on the sea floor. Clues to solve this puzzle are the absence of sedimentary bedding produced by currents, and the presence of abundant bioturbation (burrowing by organisms). There were certainly burrowing bivalves present (*Panopea*), and probably also a fauna of arthropods and worms that burrowed in the sediment. The activities of these animals were responsible for upsetting the orientation of the ammonites when they were buried by a few centimetres of sediment. In a recess in the cliff below the sandstone there is a thin bed with a concentration of belemnites, and also, as a cephalopod bonus, the remains of a nautilus. At least there used to be, until a party of geovandals tried to extract it with sledge hammers. Such destruction is especially unfortunate as the nautilus was shown to visiting field parties from many universities.

Fossil localities in Scotland have suffered the raids of inconsiderate and greedy commercial and amateur collectors for many years. Some classic fossils from protected sites have even been sold to reputable museums. Scotland now has a published Scottish Fossil Code produced by Scottish Natural Heritage. This code gives guidance on responsible fossil collecting and encouragement to amateur collectors.

At Bearreraig Bay the ammonites are attractive and well preserved, with calcite filling the inner chambers of the shell and the body chamber filled with sediment that filtered into the aperture after death and decay (or consumption) of the animal. No scientific harm is done by collecting from boulders found loose on the beach, but please do not hammer the outcrop; it is important that everybody can observe and study fossils that are still within the bedrock. After an hour or so searching the beach for fossils at Bearreraig, why not drift back in time to the Jurassic seas of Scotland and the delights that might have awaited in the shallows along the beaches, or in offshore seas.

Ellie, my research companion on Excursion 9, is an American expert on modern cephalopods, and we combined our interests with the aim of studying behaviour in Jurassic cephalopods. It was a relatively simple matter to take the Bus back 175 million years to the mid Jurassic. We could also position ourselves roughly in the area of Trotternish on Skye using palaeomagnetic data and basin reconstructions, and we needed to find some major landmarks that would help guide us to the correct spot.

We expected a large emergent 'Scottish island', covered in vegetation. In an area that is now part of the North Sea about 200 kilometres northeast of Aberdeen there should be a large dome with evidence of volcanoes. River systems carried sediment north to a delta in the region of the Brent oilfield in a great valley between what is now the Shetland Islands and Norway. To the south, in the area that is now England, there would be a shallow sea with white carbonate sands and clear blue water. Our point of interest would lie on the west side of the Scottish island, nearshore and close to an area with strong tidal currents sweeping northeast up a narrow, funnel-shaped seaway. Farther to the west of the Scottish island would lie a narrow seaway or seaways on the other side of which was 'America'. This was a time prior to the opening of the North Atlantic Ocean, and only a series of faulted rift basins, partly drowned by the sea, separated Scotland and North America.

Spotting the general pattern did not prove to be difficult. By using element mapping of the area we soon located the volcanic region of the North Sea and we could then make out the Scottish island, which clearly extended up as far as the Shetland islands. We spent a long time at orbit height and ran the time clock at a slow rate of 10,000 years/minute. We wanted a record of as much time as possible, hoping for about 20 million years of Jurassic time. This would take more than 30 hours of excursion time but was essential if we were to locate the correct time and position for the ammonite beds at Bearreraig Bay.

Once a sequence like this is run it is difficult to call a stop—the changing scene is just so wonderfully fascinating. Seas expand and contract, fingers of sea extend rapidly along active rift faults, only to retreat as sediment fills the sea from erosion of local uplifted areas. Vegetational colours change, reflecting long-term climatic variation. Time is changing too fast (166 years/second) to see any seasonal changes in vegetation, and the time from the melting of the ice sheets of the last glaciation of Scotland to the present would take only a minute.

The volume of data acquired was vast during such an exercise, providing analytical work for armies of graduate students in institutions rich enough to buy into the project. Oil companies invested vast sums in the data which they used in the analysis of oil-prone areas. This had become the new tool for sedimentology, palaeogeography and structural developments; an integration of remotely sensed geology and palaeogeography, rather than the laborious analysis of core data. As the geologist on this excursion, I had the task of finding the location of Bearreraig Bay in the *Ludwigia murchisonae* zone of the mid Jurassic. Even if a known feature (for example, a volcano) can be recognised, a mere bearing and distance need not bring us to the correct point. Crustal stretching and lateral displacement in faults are just two factors that distort the spatial relationships of surface features as the time clock tracks backwards.

We had started to record the scene within the late Jurassic, in the Oxfordian stage. Seas were extensive and sea level high at this time, and we hoped to get a good fix on a marine transgression that took place both in the Moray Firth and to the west of Scotland in the *macrocephalus* ammonite zone of the Callovian stage. Prior to this time, extensive lagoonal and terrestrial conditions covered the Outer Hebrides, represented by a group of rocks known as the Great Estuarine Group.

At this point we were delighted to see the blue of the sea retreat rapidly, to be replaced by the green of plant growth, and a rapidly shifting series of lagoons and deltas. On the west coast of the Scottish island we clearly noticed the sea retreat northwards from the Minch, thus creating a land connection between the main Scottish island and the Jurassic version of the Outer Hebridean isles, mainly represented by Lewis.

Soon we were back in time to the Bathonian stage of the Jurassic, and when the sea returned we would be down in the Bajocian—within which lies the ammonite zone of *Ludwigia murchisonae*. Only a few million years to go. To the east, land was emerging in the North Sea, and as we tracked back the heat sensors revealed intermittent volcanic activity; this was the mid-Jurassic volcanic dome in action.

Suddenly we saw the sea return to the area of the Minches, flooding the lagoons and delta plains. We were now only a few ammonite zones away from our target. In terms of the rock record we were at the rocks on which the Storr Lochs dam is built at Bearreraig Bay. Our time journey just had to track down the sandstones and shales of the cliff to bring us to the beds with *Ludwigia* (Fig 9.3). Some time in the next two hours of time-scanning we would pass through our desired time. All we had to go on was that there were three cycles of sedimentation in the Bajocian rocks beneath the Great Estuarine Series of the Bathonian stage. These can be seen as transitions from shale to sandstone in the cliffs at Bearreraig Bay. Hopefully we could spot these as minor transgressions and regressions.

It was confusing. There were far more than three shifts of coastline, but the pulses did seem to wax and wane. I clicked the cursor after about an hour and twenty minutes at what I thought might be the correct time and place; we would have to play that part slowly and carefully for more clues later.

Soon we were out of the mid Jurassic; the broad picture had been captured, and events west of the Hebrides had been recorded in detail. I ran, and re-ran, the record of the mid Jurassic from the west coast of Scotland. Ellie patiently watched and listened, offered encouragement and asked useful questions on 'why' and 'where'. Now I knew why I was on this trip; recall of details of the varied Jurassic successions of the Hebrides was essential. Eventually I could recognise the Inner Hebrides basin and the Sea of the Hebrides basin, with a fault-bounded ridge between. This gave a reasonable fix for the position of Bearreraig Bay and our ammonite beds. When we watched this area the shallowing-up cycles of sedimentation could be made out in terms of sea colour, reflecting bottom conditions and depth. Hoping I was picking the correct cycle, I tuned our time to the transition from muddy to sandy sea-bottom conditions in the cycle, for it is at that point in the sequence in Bearreraig Bay where the ammonites are found. We had made a decision; now I was on tenterhooks until the dive.

With the Bus in amphibious mode we made our landing on a calm sea. We gave each other a confidence-boosting grin and set the controls for dive, and off we went into the unknown, but frequently imagined, Jurassic wide blue yonder.

The water was clear and not particularly deep, only about twenty metres, with dappled sunlight penetrating to the sea floor. As we descended, the bottom appeared as elongate patches of light and dark, with a distinct orientation. Our current meter was indicating that we were in a current of around one metre/second and that this current was aligned with the features we could see on the sea floor. Heading into the current we descended gently. There was not a lot to see in mid-water—no teeming shoals of fish surrounding our craft—but there was movement farther away, just on the limit of vision, although shapes were not discernible.

Soon we were within a couple of metres of the sea floor, and the light and dark patches seen from above were resolved into light patches of bare sand, with ripples being driven by the current, and thickets of a seaweed standing up to fifty centimetres high. The flat, dark greenish-brown fronds had crinkled edges and were secured by holdfasts in the sediment and on to any objects on the sea floor. The nature of the objects was not immediately apparent due to the seaweed growth masking all form. Proceeding slowly close to the bottom, we scanned the seaweed and sand for signs of life, but little moved. There was nothing in sight to tell us we were in the Jurassic; this could have been seaweed-covered ground almost anywhere, almost any time.

'Hope it gets better than this,' muttered Ellie.

'So do I, but at least the report will be brief.'

'Are we at the right time?'

'Give me a chance Ellie. I know I have produced no evidence so far.'

'Sorry, no pressure.'

We pushed on, saying little. We were lying full length in the Bus, side by side and peering out of the forward observation bubble.

'Look. Shells—and some are ammonites.'

'Thank goodness for that. Something to report.'

We were over a patch of sandy sediment, where the current had removed the loose sand to reveal a variety of shells protruding from the sediment. There were bivalves, with shapes typical of those that live in burrows in muddy sand, and the shells were sticking out of the seabed in what had been the living position of the shell. More exciting was the sight of an accumulation of ammonite shells, part filled with sediment and scattered on the surface.

'I'm not sure, but they look quite like *Ludwigia*, don't they?'

'Yes, but the colour pattern is confusing.'

Many of these shells although dead still had colour, and the colour patterns dominated the sculpture of ribbing, giving them an unfamiliar look in comparison with the fossil version. Although fixed automatic wide-angle video cameras were recording both lateral and forward views, we still needed to take stills of interesting features. The camera was external to the Bus with remote controls, Ellie had to control the position of our craft, and I had the task of pointing the camera using a remote viewfinder with x–y shift sticks in the Bus. Eventually I reckoned we had some good shots, with ammonite shells lying on the sea floor.

We were about to explore further when a cloud of sand erupted from a hollow beneath an ammonite shell. Eruptions continued until the rear end of a burrowing shrimp emerged from the hole, only to disappear again rapidly down the burrow. It was using the shell as a shelter for a burrow entrance, and as it burrowed under the shell it subsided into the sediment—the first stage of burial in its long journey to becoming a fossil.

As Ellie increased thrust for us to move off once more, the rear of the Bus dipped and gently bumped the seabed. There was no need to worry, and the result was interesting. Several streaks of iridescent blue and pink shot past the window, one leaving a dark inky cloud in its wake.

'They must be belemnites.'

'I suppose we sat on them and they objected,' I suggested.

Certainly we had disturbed them when we bumped the bottom. They must have been resting or hiding in the seaweed, or merely have been camouflaged against the sand and seaweed, and we could not see them amongst the underwater reflections. Since the Bus did not give chase, the belemnites rapidly slowed and we were soon in close contact with them. They now swam gently in a close group like a small shoal of fish. They were on alert, swimming backwards, and the pulsating opening and shutting of the hyponome revealed that they were using their 'jet propulsion' mechanism, but in low gear. To achieve this type of propulsion the belemnite draws water into the mantle cavity under low pressure and then squirts it out of the narrow flexible opening of the hyponome. It is the same principle as a water pump or jet engine. Modern squid and octopus employ the same propulsion technique, shooting backwards to escape enemies, and in many types they produce ink to lay a 'smokescreen' to confuse an attacker. When foraging, or swimming slowly, cuttlefish move forwards by undulating the flaps of skin ('fins') on either side of the body.

The belemnites were still wary of the Bus and were not in the mood to show us this behaviour. One thing they did do was rapidly alter colour to match the background, changing from a brown mottled sandy colour when swimming over sand to green and black when over the seaweed. They employed and therefore presumably required camouflage—but against what?

'Oh, look at that, will you.'

'Where?'

Ellie's finger stabbed in front of my vision pointing to our lower left.

'Oh, marvellous, fantastic, what a beast,' I enthused.

'Yeah, really cool, and beautiful.'

I had to agree. Swimming sedately past with a slight rolling motion—or was it a natural sexy wiggle—was a fine ammonite. It was some twenty centimetres in diameter, pale bluish with lighter mottling on the underside, but with bold brown and green bars over the top part of the shell. She—for it must have been a female *Ludwigia* because of its size and rib pattern—had large round eyes with prominent pupils, and her tentacles were bunched together in a manner that perfectly balanced the shell. Above the tentacles a hood of flesh merged with the shell outline to complete the streamlined form. This ammonite seemed on a mission with purpose, doggedly plodding along using her 'jet propulsion' capability in the same way as the belemnites (Fig 9.6).

'Follow her,' I urged. 'How fast is she going?'

Ellie gradually swung the Bus around, and we gave chase at a respectful distance. The ammonite was doing about three kilometres an hour, but quite suddenly she seemed to associate our Bus with danger and descended quite sharply till close above the dense seaweed growth, where she slowed down. Quickly she spread her tentacles wide, inflated the mantle and in a flash shot into the seaweed and out of sight.

'Oh, bother, she's gone and hidden. She must have thought us to be a large predator, and certainly did not want to be followed.'

We hovered near the spot for a minute or so, but there was no sign of the ammonite, so we moved on.

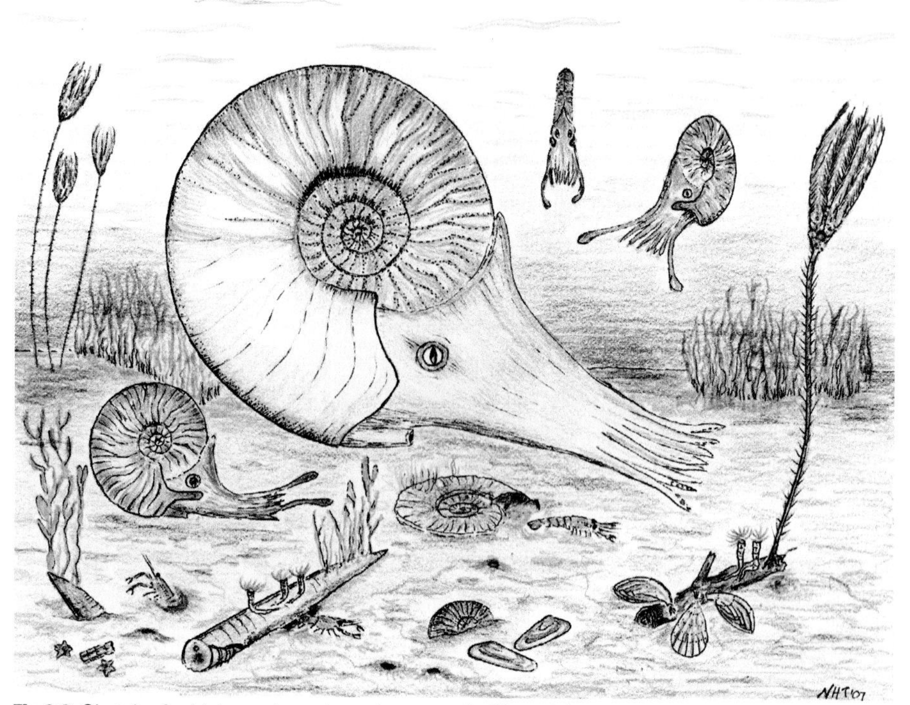

Fig 9.6 Sketch of mid-Jurassic underwater scene in Skye. A female *Ludwigia* is courted by three smaller males. On the sea floor a waterlogged piece of wood provides a holdfast for a crinoid, worm tubes and brachiopods. Belemnites, ammonite shells, bivalves and crinoid stem fragments are partly buried in the muddy sand, and shrimps emerge from their burrows to feed on the surface.

'There's another, over there—and another to the left. It's odd that they're all going the same way.'

Both of these ammonites also dived into the weed when they felt we were too close. We were now better at spotting them, and realised that we were causing some panic in the world of the cephalopods. They were certainly wary of large moving objects.

The lack of fish seemed surprising. Were they just hiding, or were there no shoals of mid-water swimmers in the area? Maybe they were also frightened by the size of the Bus. The odd fish did dart off from some sand patches as we passed, but we saw nothing exciting or clearly identifiable. Fish fossils are not common in the mid-Jurassic rocks of Skye, even as scattered scales, so there is little evidence to use from the rock record at this locality. What we did spot were a few small shoals of fishes with shiny rhomboid scales. Being not more than twenty centimetres in length, they looked superficially like herring. There was also the occasional small shark, which were no different from modern small bottom-living sharks.

'What was that?' Ellie queried.

We both felt it—a distinct bump on the Bus—yet we were well off the bottom, and the bump was from the rear. The stability was briefly upset, and we wobbled a bit before regaining our steady, stable progress. Ellie started checking instruments, and we exchanged questioning glances; adrenalin levels rose.

A long shadow moved over the seabed. Something large had joined us. Maybe it saw us as competition. After the shadow had passed, there was another wobble of the Bus. Whatever it was, it was large enough to create a wash that could push us about.

'Wow, there it is.'

Suddenly, right in front of the window was the bulk of a large animal. Its leathery-looking skin was dark grey and striped with lighter bands, and there were clear scars where it had been injured. As it moved away, a giant flipper came into view and swept majestically past the window, giving us another rocking. It was turning, and in a flash a snake-like neck with a small head appeared in our sight.

'Plesiosaur,' we chorused.

It was big, and we were clearly being inspected rather closely—rather too closely for comfort, in fact.

'Could it do us damage?' queried Ellie.

'Well, I'm not going out to swim with it. Do you see those teeth?'

'Sure do; no wonder the fish and ammonites are a bit nervous; he looks mean.'

'It could be a she,' I suggested. 'They can be mean as well.'

'Here it comes again.'

This time it circled us, giving us a fine display. It was a highly manoeuvrable beast, with four large paddles and a long neck.

'I think he is confused. The reptile brain will not compute this experience for him.'

'Maybe he can't decide whether to attack or try and mate with the Bus.'

'I think he is losing interest. He seems to be ignoring us a bit more.'

Suddenly the plesiosaur turned very fast, and thrust his head, mouth agape, into the weeds. Belemnites streaked out in all directions, but one was left clamped in the vicious jaws. The water was black with ink clouds, but our plesiosaur gyrated rapidly, neck striking out, and he bagged three victims before the rest found a new hiding place. At this point he lost interest in our strange craft and cruised off into the distance, continuing his killer patrol. We drifted on our way seriously impressed by our encounter with one of the top predators of the Jurassic seas. Even the most cynical student would give this a rating of 'very cool'.

Ellie was chattering excitedly as we drifted, speculating on whether the plesiosaur also liked ammonites, when we realised we were putting ammonites down into the seaweed all around us. We were certainly influencing the environment, so we needed to sit and watch rather than cruise. With the motors shut down we came to a gentle stop, resting lightly on a patch of sandy seabed near a spot where several ammonites had disappeared into the weed. The landing was gentle but still stirred up some of the sediment.

The current seemed to be minimal now, and the sediment cloud drifted slowly, with a slight rocking motion, indicating we were within wave-base. With the noises of motors abating, a gentle stillness enveloped the Bus. Outside it the marine life seemed to be relaxing. A couple of large female ammonites were swimming slowly at the edge of the sand patch, their mantle cavities gently pulsating and coloured a pale pink.

'I didn't notice that colour before. What are they up to?'

We soon had an answer, as several small ammonites appeared from the weed and approached the females.

'Oh, fantastic—they are mature males. What little beauties, and look at the lappets.'

Sure enough they had lappets and they were a brilliant red colour—a pretty obvious signal of intent. They also had two tentacles longer than the others, and these they waved alluringly in sight of the females. The males seemed cautious; maybe the alternative to mating was to become the evening meal.

The ammonites continued their courtship dance, and we watched the male gently approaching the female from behind and slipping his tentacle inside her mantle cavity, where he deposited his packets of sperm to fertilise the eggs she carried. At the point of contact the female was suffused in a pink glow, and red colours in the mantle of the male positively throbbed. We followed the rituals of the ammonites closely as the Bus rocked gently on the Jurassic sea floor.

'I think we have cracked it, that is exactly what I hoped to see.'

'Glad you are happy, Ellie. I am mainly relieved that we arrived at the right place at the correct time. All the rest is a bonus.'

The rocking of the craft stirred up more sediment, and somehow we had failed to notice that we had attracted a number of fish that were busy probing the sediment for small worms and crustaceans. We were about to call 'time' when a nautilus suddenly appeared alongside the Bus. It was interested in the disturbance we had made to the seabed, and was hunting for food with a cunning technique. It would spread its tentacles and then blow away the sediment with blasts of water from the hyponome. Anything that moved was then grabbed by the tentacles. It had a near miss as a large shrimp shot from its clutches, but the next one was not so lucky. Snared at first on a single tentacle it was close to freedom, but once other tentacles made contact it was dragged into the beak and consumed. This nautilus was a messy eater, and small fish gathered to pick up the scraps, giving us a good photo opportunity and the possibility for identification of the fish from the photos at a later date.

We now had a perfect Jurassic scene of tranquillity with feeding fish and well-satisfied ammonites, nautilus and researchers. As with any wildlife watching, the key was to keep still and wait for the creatures to relax and adopt the normal behaviour of feeding, procreation and house-building. Whilst it was moving around, the Bus had been threatening to most animals we had seen, and all we had observed were fright and flight reactions, and maybe threats from the large plesiosaur.

After an hour of observation we had a much better idea of the variety of fauna, and several animals such as the nautilus were adopting the Bus as a new territory. We made a significant feature on the sea floor, providing opportunities for both shelter and ambush, in much the same way that a modern shipwreck forms the physical environment in which an ecosystem new to the area can flourish.

Reluctantly we gently started the motors and lifted from the sea floor. The muddy sand we stirred up contained a minor feast for the fish and nautilus, but they would have to find a new shelter. Back on the surface we paused to programme the Bus for the journey back to the present and a dry landing.

Back in the laboratory we spent the next few weeks reviewing the data we had collected, particularly examining all the film and photos. As with any expedition of this kind we had pictures of many animals that could not be identified, particularly soft-bodied creatures such as worms and jellyfish that are seldom fossilised. But we also had positive identifications for ammonites and other creatures with shells based on the fossils that can be found at Bearreraig Bay on Skye. Now we could combine our observations with the known fossil record to produce a better understanding of this Jurassic marine environment.

As with all fieldwork, we wanted to go back and have another look. There are always questions that arise from data gathered on a first visit that require a second trip to provide an answer. This is how research workers become 'experts' on a particular area, and may even display territorial tendencies, not welcoming others to 'their' area. We had no opportunity to return to the past; the grant money was all spent. However, we could go to Bearreraig Bay and see the evidence in the rocks.

Thus in the spring of the following year, before the Skye midges emerged, Ellie the expert in modern cephalopods was introduced to the study of fossils in the field, and to the social life of the western Highlands.

Excursion 10

The Helmsdale Tsunami

TIME: Late Jurassic.

LOCATION: Helmsdale, Sutherland.

OBJECTIVES: To examine an active submarine fault and study the effect of the fault on the landscape, and to research the wildlife.

THE MODERN EVIDENCE: Boulder beds banked against a major geological fault that defines the coastal lowlands from Golspie to Helmsdale. Fossils of wood, corals, ammonites, crocodile and ichthyosaur from the beach at Helmsdale.

Situated on the northwest side of the Moray Firth at the mouth of the Helmsdale river is Helmsdale itself, which boasts an old bridge and coaching inn established in 1812 that some years ago self-advertised as the 'best hotel in the world'. The hotel operated at the cheap and very cheerful end of the market, and in the winter bed-and-breakfast at £5 a night was available for student excursions. However, despite healthy bar takings, the hotel shut down. Now, after impressive renovations and new plumbing it is open again, with excellent food and still catering for geologists. There is a heritage centre at Helmsdale which is well worth a visit, together with several other hotels, guesthouses and bars. Most souls that reach this corner of Scotland flash through the town over the 'new' bridge en route to the tourist shrine of John o' Groats, which is not even the most northerly mainland point. For the most northerly point it is necessary to take the minor road to the tip of Dunnet Head. On a clear day the view is excellent, and it is possible to see Orkney as well as cliffs of Upper Old Red Sandstone—the same age as the rocks of Dura Den (Excursion 4). Very rare scales of *Holoptychius* have been found near Dunnet.

Attractions in Helmsdale include a touring theatre, a folk festival and fishing in the famous Helmsdale river for salmon. 'Geotourism' also brings many people to Helmsdale every year. For others it is the lure of gold and panning in the Kildonan burn at Baille an Or for the scraps left after the Helmsdale goldrush of 1869. Some take this very seriously, but it is not a route to riches, and if expenses can be covered the geotourist is either very clever or more likely a blatant liar.

Fig 10.1 View of the coast north of Helmsdale to Dun Glas (Green Table). Upper Jurassic Helmsdale Boulder Beds are exposed on the shore. The cliff to the left is Helmsdale Granite and the Helmsdale Fault crosses the beach and passes through the dip at the back of Dun Glas and continues out to sea to the northeast.

Large numbers of geologists visit this area to marvel at the exposures of the Helmsdale Fault and the Helmsdale Boulder Beds (Fig 10.1). The geological map shows that this fault defines the coastline from Golspie to Wick, and has on its landward side the metamorphic and granitic rocks of the Highlands together with deposits of the Old Red Sandstone. However, a low-lying coastal strip to the east of the fault from Golspie to north of Helmsdale consists of younger Jurassic rocks, which yield a variety of fossils to the patient observer.

In the Brora area a coal seam was worked intermittently from at least 1592 to 1972, and one of the most celebrated geologists of the nineteenth century, Sir Roderick Impey Murchison, published an account of the coalfield at Brora in 1827. The coal was used as house coal, and to fire the kilns of the brickworks, which utilised local clay. In earlier times the coal was used to provide heat to evaporate seawater in salt production, and there is still a Salt Street near the sea on the south side of the Brora river. Thus the local geology has played an important role in local industrial history; this was the most northerly outpost of the coal-fired industrial revolution.

To the north of Brora, from Kintradwell to Helmsdale there is a feature unique in British onshore geology. This is an example of a fault that cut the Jurassic

Fig 10.2 Surface of a typical Upper Jurassic Helmsdale boulder bed with clasts of Old Red Sandstone. The bed was deposited by a debris flow from the submarine scarp of the Helmsdale Fault. Debris flow was probably triggered by an earthquake associated with movement on the Helmsdale Fault. Photo taken on the beach north of Helmsdale harbour wall.

seabed 150 million years ago and separated a Scottish landmass with a sea beach and a narrow shallow marine shelf to the northwest from deep-water conditions to the southeast. The presence of a fault is not remarkable, but what is unusual is to find preserved in great detail the rocks, sand and organic debris which tumbled off the exposed submarine fault scarp to be deposited in deep water.

Over most of the fifteen-kilometre length of exposure of the fault, the rocks in a strip up to a few hundred metres in width on the downthrown side of the fault consist of rockfall breccias. Tongues of material extend farther from the fault as boulder beds—beds with pebble and boulder-sized rock fragments in a matrix of sand or broken debris of shells (Fig 10.2). At low tide these rocks can easily be seen to the north of the harbour wall at Helmsdale. Other deposits include dark shales, often with thin, centimetre-scale beds of sandstone. These shales, sandstones and boulder beds contain fossil land-plant material, including trunk fragments of Jurassic pine trees, and also the remains of marine animals such as oysters, corals, ammonites and belemnites (Fig 10.3). The rocks must have originated in marine conditions, but land was not far away, as is illustrated by the plant remains present.

Fig 10.3 A: Colony, 25 cm across, of the compound coral *Isastraea*. **B:** Detail of the surface showing the individual, tightly packed corallites, each about 5 mm across. Specimen from the Helmsdale Boulder Beds, Sutherland.

So what was going on 150 million years ago in Helmsdale? Our aim was to explore the beach and coastal features, and relate our observations to the rocks and fossils seen at Helmsdale today. Jeremy—my companion from the Caithness Devonian fishing excursion—and I were ready in the Bus, with our landing area on the beach programmed into the controls. We were off, and minutes later the Bus made a perfect landing.

We emerged on to a sandy beach in bright sunshine. On the face of it this seemed a reasonable beach. There was a small cliff forming the front of what appeared to be a flat terrace. It was covered in pines and ferns, and plants grew from the cliff face and along its base. Clearly the high tide did not reach this cliff (Fig 10.4).

Jeremy and I were standing on a beach with a strip of pebbles all of pea to marble size. They glistened in the sun and had been well rounded by the pounding of the waves. Farther down the shore the beach was sandier and contained lots of shells washed to and fro in the surf. Most were broken and worn, but there were many oyster fragments and pieces of gastropods, bits of coral and spines from sea urchins.

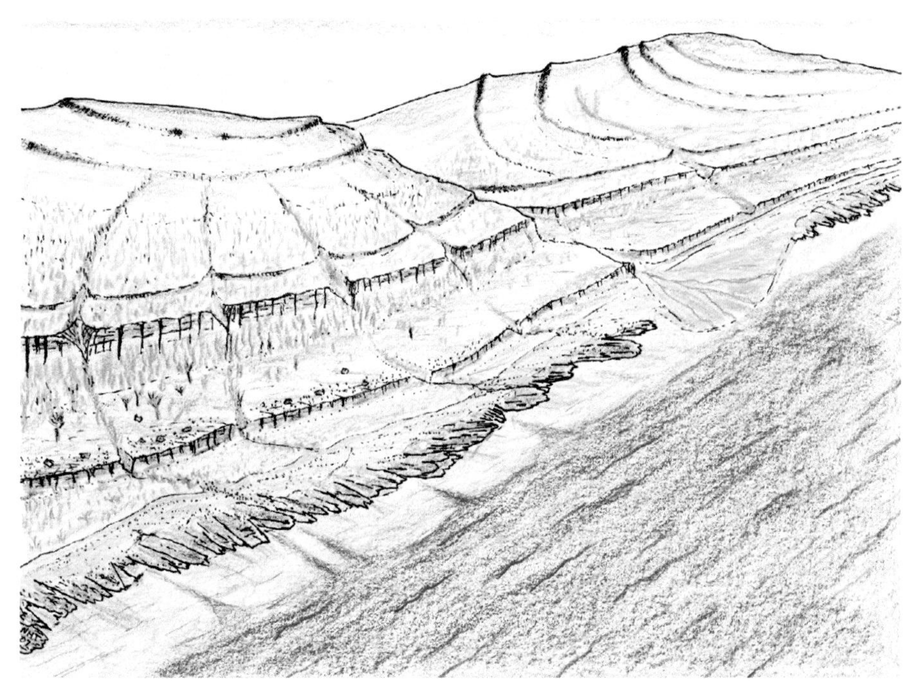

Fig 10.4 Sketch of the Helmsdale shore in the late Jurassic. Raised marine terraces are cut in Middle Old Red Sandstone flagstones which also appear in reefs on the shore. A small delta covers part of the shelf. Offshore the rapid change from light to dark blue marks the submarine position of the Helmsdale Fault.

The lower shore was largely rock with some loose material stuck in gullies and hollows. Oysters encrusted rocks that were also bored by bivalves. Sea urchins grazed the rock surfaces, scraping off the algae, and brachiopods clung by their pedicles in the crevices. Poking in crevices (Fig 10.5) and under stones in the pools produced some surprises: small fish, shrimps and crabs were disturbed from their homes, and a great variety of small gastropods were hiding under the larger stones. This is just the situation on rocky shores today; it is necessary to look under the rocks to reveal the variety of life in the intertidal area.

One hole under a large flat rock in a deeper rock pool had a pile of broken shells at the entrance. A poke with a stick produced an angry reaction, and strong tentacles grabbed the stick. This was the home of an octopus. It let go when I tried to pull it from its lair, so we left it in peace. In the larger hollows on the rocky platform the sand was white and made almost exclusively of shell debris.

We stood up to our knees in water looking out to sea. The sea was calm, and the waves only splashed up to our thighs; it was warm, and as if in confirmation, coral colonies were dotted around cemented to the rocks. The coral heads nearly poked out of the water, so it must have been fairly low tide—corals die if exposed. This rocky shelf had many familiar features. Apart from the fact that the species of animals were not the same, the general environment and fauna reminded us

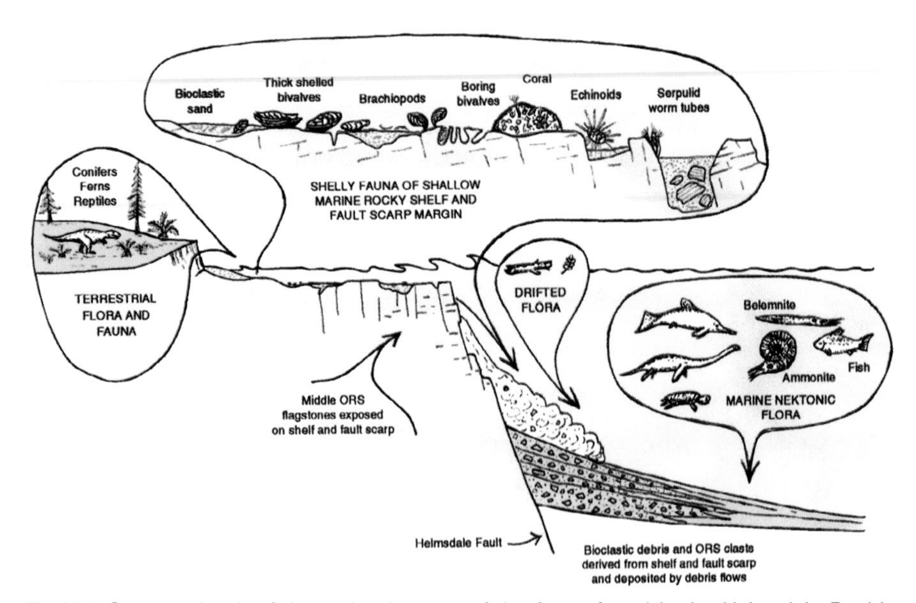

Fig 10.5 Cartoon sketch of the main elements of the fauna found in the Helmsdale Boulder Beds. Plants were derived from the land, and shelly fauna from the rocky marine shelf; and they are preserved in the boulder beds and shales along with an open marine fauna dominated by reptiles, fish, ammonites and belemnites.

of rocky reefs just south of the tropics in Western Australia, where we had fished for anything that might show up—apart from blowfish. As fishermen rather than as geologists we stood there staring out to sea, and scanning the waves and water for signs of life.

'Something splashed at the surface just beyond the breakers. Yes, there is a long dark shape out there.'

'Come on, Jeremy, you are obsessed with fish—it's a log.'

'But logs don't swim. It's moving. See, it's turning round and coming this way.'

At that point a head appeared—not much of it—but it snorted and sank.

'That's a crocodile. Let's just go back to the beach, shall we?'

We rapidly retreated. Saltwater crocodiles have a bad reputation today, and we did not need to check the interpersonal skills of a Jurassic version. A more considered view from the beach revealed that our crocodile was much closer than we thought, but also much smaller and not a serious threat. It was another Loch Ness Monster scenario; if the expectation is to see something large the brain obliges with that interpretation, so a diving otter can become monster-sized.

We wandered along the beach leaving distressingly humanoid footprints for the late Jurassic. With the wind at our backs we nearly stepped in a mass of rotting flesh and bones on the strandline. The smell was that pungent, gut-wrenching aroma typical of dead sea animals, yet this could not be seal, dolphin or whale, and certainly was not a sheep fallen over the cliff.

'Marine reptile, pretty bad order and odour. I'm not trying to collect that,' I ventured.

'You can't anyway. It's against the rules,' retorted Jeremy.

'Okay, just a joke. But I can't find its head. I think it was washed up dead, and the head has fallen off or been bitten off.'

The flesh had rotted badly, and bones of the rib cage protruded from the dried surface of the skin. A long neck revealed elongated vertebrae with strong tendons. A large paddle emerged from the shoulder girdle, and the bone pattern showed through the shrunken leathery skin. There were signs that the carcass had been scavenged, but it was now too far gone to be tasty, even to a scavenging dinosaur. Although the area around the carcass above the tide line was rather disturbed, there were no clear footprints in the sand. Parts of the carcass were either missing or buried in the sand, and we had no stomach to start digging. Maybe it was already incomplete when washed ashore, after the crocodile had possibly had a meal or two. Anyway, it was clearly a form of plesiosaur, and although we could not see the head and the teeth it probably fed on fish and squid-like animals such as belemnites.

We were about to depart the scene when we noticed some tracks. The sand was smooth on the seaward side of the carcass where it had been washed by the last high tide. A different story was revealed on the landward side, where there was a lot of disturbed sand and some distinct, three-toed footprints—about the size a heron would leave; but we had seen no birds, and maybe there were no birds to run on sand at this time. These must be prints of small bipedal dinosaurs; Jurassic Park came to mind, and we both paid more attention to the landward view rather than gazing out to sea.

We left the smelly brute and went to inspect the fern-covered cliff at the back of the beach, while at the same time keeping a lookout for hungry little reptiles. Climbing up on to the cliff terrace was not easily done, even though the cliff was only four or so metres high. It was strongly overhanging, with an undercut notch at the base, and any cracks in the cliff were colonised by thick ferns and small conifers.

'This coastline is rising—that notch is due to bioerosion more than physical processes. Look, you can still see holes left by boring bivalves, and some chalky old oyster shells are stuck on the surface. The sea no longer reaches this cliff, so the land here is rising relative to sea level. I bet the shelf above the cliff is a raised beach.'

Sure enough, when we eventually found a way up a gully in the small cliff we emerged on a gently sloping platform, at the back of which was another cliff. This was more degraded and vegetated than the one we had just climbed. Above that we could just make out the line of a third terrace through the vegetation. Large rocks were also scattered on the surface in a random manner—a strange feature since they did not look as though they could have tumbled and rolled to their resting places from the higher cliff.

'Right. The way I perceive it is that the land is rising, but the gentle uplift to give a sloping terrace is sometimes interrupted by sudden change.'

'How about sudden changes in sea level, like those in the ice ages?' suggested Jeremy.

'Possible, but I don't think we have any ice caps in the late Jurassic, although sea-level changes did take place anyway.'

'How?'

'Depends who you want to believe,' I replied, 'but I think we should investigate some more.'

'Not too far uphill, please. I don't feel like a climb or an encounter with any of the larger fauna that might be lurking in the woods.'

'It's all right. This spot will do. If you look at the shore and the rocky shelf with the waves, what strikes you?'

'The shelf is narrow, the water colour changes dramatically to dark blue just beyond the breakers. There must be a sudden drop off, we will have to look, but after lunch.'

From the terrace edge we had a view up and down the coast (Fig 10.4). The shore was virtually a straight line on the large scale and clearly controlled by the Helmsdale Fault. It was low tide, and the sandy beach below our vantage point changed to rocky reefs lower down the shore. This seemed to be the norm along most of the coast we could see. The reefs were made of Old Red Sandstone flagstones and sandstones, dipping at gentle angles. The only interruption to this scene was the small delta a mile or so to the northeast, which covered the beach and foreshore and allowed vegetation to extend closer to the sea. It looked as though longshore drift was to the northeast, since the delta was deflected along the shore in that direction. The delta was fed by a small river that emerged from a valley that cut through the cliff and shelf features. The valley seemed to be exploiting a weakness caused by a fault that cut the Old Red Sandstone. We could see that the dip and strike of the flagstones changed north of the valley, resulting in a different orientation of the rocky reefs on the foreshore, and ridges on the hillsides marked the positions of more resistant sandstone beds.

We had been too hasty in leaving the Bus and had left our lunch behind, so it was time to go back and plan the rest of the day. Fortunately we were able to walk fairly easily along the edge of the lower terrace—the vegetation, mainly ferns, being generally no more than waist high. There were also small conifers and cycads dotted about the shelf, many growing from crevices and hollows where thin soil had accumulated. A lot of the time we were walking on rock—dark flagstones and sandstone beds with a strong joint pattern. Many joints were roughly parallel to the cliff edge and provided the crevices for plants to take root. There were also the scattered boulders; these too were made of the flagstones. A large coral colony was trapped under one boulder, which provided more evidence for marine uplift. After a few hundred metres we found a convenient path that went down a gully to the beach.

'Wait, this is not a sheep track,' I cautioned. 'What do you think has been down here before?'

'Reptiles, Nigel, on a regular route. There are claw scratches on that rock.'

Sure enough, the side of a flagstone ledge that crossed the track was worn smooth with numerous scratches from sharp claws.

'This track must be regularly used by a lot of small animals. The marks are similar to those made by the strong claws of rockhopper penguins on sandstone ledges along their cliff paths to the breeding colonies on the Falklands.'

Taking no chances, we inspected the gully from above, lobbed rocks into the vegetation beside the track, and generally made a lot of noise. It was just such a good place for an ambush. A large predator could lurk in the vegetation and grab one of a group as they came down to the beach. We descended the gully with caution, acutely aware of any rustling from the ferns beside the path. We leapt back and then laughed as a lizard shot across the path, but that was the only alarm.

We were soon on the beach, and there were the same three-toed footprints we had seen beside the plesiosaur carcass. But there were also bones here—a few vertebrae, a limb bone and a lower jaw, about ten centimetres long, armed with a row of sharp, dagger-shaped teeth. Then there was a dried tail in mottled brown with dark red above, with a pale underside, and attached to a hip bone.

'I think we were wise to consider the ambush possibility. Something big has been feeding on these beasts. Look at the damage on this bone.'

'Imagine the tooth or claw that made that hole,' shuddered Jeremy.

As we walked back to the Bus along the beach, keeping well away from the vegetation at the top of the beach, we imagined the scenario if we met a real live dinosaur. The odd dinosaur bone and footprint have been found in Skye, so they were on the Scottish island and could be behind us right now, eyeing up a mammalian virtual meal. We ate lunch in the Bus and then returned to the beach to inspect the edge of the shelf. By then we had not seen the crocodile for more than an hour. However, there could be plesiosaurs and ichthyosaurs out there, but they should be more interested in fish than in us.

'I have seen a spot where it is really shallow out to the edge of a gully, and I am going to snorkel out and look at the edge.'

For this venture I used a wetsuit, not only to protect me against a nasty poisonous bite or sting but also to avoid being cut on rocks or coral in the gullies. I also donned flippers and a snorkel mask. Jeremy was better on the computers in the Bus should anything significant happen to me in the water.

When sploshing into the gully, the first problem was balance and orientation. The waves surged up and down past walls encrusted with grazing sea urchins in gaudy reds and blues. Coral colonies, but of only one kind, clung to the edges of the gullies, and fish darted into crevices in the rock walls where joints and bedding planes in the flagstone walls had been opened by the force of the waves. The floor of the gully was covered with quartz pebbles like those from the beach, together with broken shell debris. The beach-derived material was being channelled into deeper water down the gullies.

As the gully deepened the wave surge was easier to control, and soon the blackness of the dropoff loomed ahead. Suddenly I was in space—no rocks, no sea floor in sight—and experiencing feelings of vertigo. Turning round I stared at the dropoff. It was almost vertical and stepped where blocks of rock had been released along joints and bedding planes in the flagstones. Sand and shell fragments draped the ledges, but the dropoff disappeared into inky darkness. Farther along, a cloud of sediment suddenly broke free and a rock tumbled into the depths. It was strange that I had not seen anything move it. Maybe it was just a fish or crab digging in a crevice. A few fish floated along the edge of the dropoff—some of a decidedly catchable size. To my left a shoal of fish near the surface flashed in the sunlight flooding through the clear water. A large splash nearby instinctively made me put my head up.

Looking shoreward I could see Jeremy frantically waving and shouting and pointing. Following his gaze I could see nothing untoward. Better go back and find out what he wanted. Before I had a chance to set off back to the shore, the water in front of me erupted in a shower of spray and the unmistakable shape of an ichthyosaur did a perfect porpoise impression (or should it be the other way round?). Ducking under, I could see it cruising past, looking curious but not cross, and fixing me with a large reptilian eye. That narrow toothy bill had to be for fish, not fat stuff like me—just trust the interpretation! I knew they liked squid, so hopefully I did not look much like one. My heart rate was well up, but after a few more circuits the ichthyosaur cruised off, and I returned up the gully trying to appear nonchalant. But it cannot have been very convincing.

Jeremy was as excited as usual. He had heard animals making strange bellowing noises and crashing through the vegetation. Something had frightened them, and they sounded quite big enough to avoid meeting. We leant against the Bus while I removed the snorkelling gear and dried off in the sun. We had only an hour of daylight left, so nothing too adventurous could be done—certainly no stalking of large animals in the bush.

Suddenly we looked at each other, eyes wide open. We had both felt it then. The Bus had quivered, shaken gently but distinctively. The door was the other side. What else was on the other side? None of these thought processes had to be spoken, obvious possibilities flashing through the mind. Within seconds it happened again, but this time a twig fell from a bush and simultaneously dry sand ran into our footprints around the Bus. The next moment there was a roar, we were thrown off our feet, and the Bus rocked and nearly fell. Panic!

'Get in—let's get out of here fast.'

As the Bus door slammed I had a glimpse of the shelf—there was no water, it was fully exposed.

'Water goes—water comes back—get this thing to 500 metres smartish and let's watch.'

The Bus started when sworn at in Welsh, and we jolted aloft. The monitor screen showed a fearsomely high wall of water heading for the now-exposed dropoff. It crashed against the face, driving blocks of rock many tonnes in weight over the shelf. The wave continued up the beach and into the vegetated cliff. It surged to the cliff top, tongues spilling on to the terrace above. Then it receded, dragging rocks, plants, animals, anything in its path, back across the rocky shelf, over the fault scarp and down to the depths.

'Well, Jeremy, if you want to see a debris flow you should be down where I was snorkelling an hour ago.'

'That would be marvellous, amazing; but for most people who observe debris flows first hand the experience is life-threatening.'

The sea sloshed back and forth on the shelf with diminishing ferocity, but when all had settled down there were changes to the scene that had originally greeted us. Large rocks littered the shelf, smashed tree trunks littered the beach and the sea no longer reached as far up the shelving beach. Oysters lay waterless under the sun, doomed to die and rot—their shells to contribute to the sand grains of the shelf.

What had happened offshore from the fault scarp—the dropoff over which I swam? There was no going back to see, but the sea was dark, dirty and opaque with suspended sediment. Well, I expect that another boulder bed had been deposited. Rocks from the fault scarp, pebbles, coral colonies and shelly debris from the shelf were all swept into deep water by the earthquake shocks and Helmsdale tsunami. This was a dramatic origin for just one of the numerous beds exposed along the shore there.

'Marvellous,' said Jeremy, 'but we did not see it all—what about those thin sandstone beds in the shales between the boulder beds at Helmsdale? If the boulder beds are due to earthquakes and tsunamis, what about all the thin sand beds?'

'Good point, Jeremy. There are tens to hundreds of the thin sandstone beds for every boulder bed, so the events causing them must have been much more frequent than the major earthquakes on the fault.'

'How about storms?'

'Yes, that seems reasonable. Storms are orders of magnitude more frequent than earthquakes, and they could stir up the sands on the shelf and create thin beds when the sand is swept over the shelf edge and settles on the sea floor in deep water. That can produce the tiger-stripe rocks with black shales alternating with thin brown sands.'

Jeremy agreed, and summed up: 'Periods of extreme violence by earthquakes, lesser violence by severe storms and long periods of normal weather conditions resulting in shale deposition in deep water.'

All the evidence can be seen in the rocks on the shore at Helmsdale. If lucky, it is also possible to find fossil coral and pine wood; if very fortunate the discovery might be a bone of crocodile, plesiosaur, ichthyosaur or turtle—all these have been found at Helmsdale. To date there have been no discoveries of dinosaurs, but it is quite possible that such a find could be made in the future—the unfortunate animal having been swept out to sea in a tsunami and drowned.

Epilogue

Our journeys are over, and we are once again in the present day. I hope the reader has enjoyed the excursions back into the geological past, and may now regard fossils as something more than a dusty relic of the past. To me it is still surprising and remarkable to discover not how different, but how similar, fossil ecosystems were to those we see on Earth today. Even 410 million years ago at Rhynie we can recognise many types of animals that still inhabit our gardens today: harvestman spiders, centipedes, mites, springtails and nematode worms, to name a few. The vegetation then was very primitive, but fungi and bacteria were already playing the same roles as they do today.

In the Jurassic, life was certainly more exciting. Dinosaurs did roam in Scotland, where they were the largest herbivores and the top predators, so naturally exciting the greatest public interest. I suspect that many students started their interest in fossils through dinosaurs, and have gone on to study geology. However, there is much interest to be found in the more general faunas and floras that have inhabited Earth, and the ecosystems they created. Again, there are so many comparisons that can be made between past and present. I am sure the delight experienced in watching dolphins today could have been matched by watching ichthyosaurs in the Jurassic seas.

Palaeontology is first and foremost a field science, and there is no substitute for the interest that is generated by collecting and identifying fossils and forming a properly assembled collection. Amateurs have made excellent contributions to the understanding of Scottish fossils and localities in recent years. However, collecting must be done with appropriate permission, and care needs to be taken not to cause damage to Scotland's fossil heritage. For those that wish to collect fossils, guidance can be found in the Scottish Fossil Code. Be responsible, follow the code and discuss any finds with experts in museums and universities.

Visiting the past provides a perspective on the present and gives hints on the future of life on Earth. Life has altered incredibly through the last 500 million years, and changes continue today at an extraordinary rate, owing in no small measure to the activities of mankind. Mankind is just one of the millions of species that inhabit Earth, but the one that is able to consciously influence the environment and to spread species from continent to continent for pleasure or economic gain. Thus mankind is rapidly mixing the ecosystems on Earth in a way that nature cannot do.

Sadly, we make big mistakes! Cane toads, foxes, cats and rabbits have been introduced into Australia; and grey squirrels, mink, rhododendrons, bracken, and giant hogweed reached Britain. Species are becoming extinct, and, given the power of the communication systems that have been developed, the whole world hears about it. We can help, and should help, to 'save' iconic animals such as the tiger, rhino, albatross and whale, but species change on a large scale is as inevitable now as it has been in the geological past—and will be in the future. Mankind is certainly having a rapid, short-term influence (geologically speaking) on the biota of our planet, and possibly we are living through a mass extinction as important as those that took place many times in the geological past.

We like to conserve things as they are, or as they were, so we manage nature reserves and national parks and try to 'preserve nature' as we imagine it should always be. This is not the way nature works, and it is not sustainable in the long term. Climate has always changed—and always will be doing so. Continents will continue to be on the move, and ocean currents will change. Nature has to move on or adapt and evolve; it has never stood still on a geological timescale. Ice from the last ice age in Scotland melted only 12–15,000 years ago—a mere blink of the geological eye. Cold-climate faunas and flora have retreated to the north with the ice, and temperate faunas and floras have arrived from the south since that time. Some animals, like the mammoth and giant elk, have become extinct relatively recently. The migrating biota has no respect for national boundaries. Thus as climate warms we hear concern from nationalistic ornithologists that the dotterel may abandon the high tops of the Cairngorms as a nesting area. However, there is no hard evidence that an overall temperature change will significantly alter a mountain-top environment; many other factors are also relevant. If the dotterel does not like the new conditions, it can join others of its kind in Scandinavia. There would be no loss to the world biota, only a sensible response by one species to changing conditions.

The idea that we can maintain biotas in a stable state by creating reserves is only tenable for a few generations but not in the long term. This timespan is excessive for politicians, and probably sufficient for the perception of most people who 'want it to be there for my grandchildren'. I am not advocating doom and gloom; I have faith in the ability of the biota on Earth to evolve and adapt to different conditions. Quite probably it will not be mankind that lives on; species that expand rapidly frequently crash just as fast. Would that really be a disaster for Earth? Mankind is only one species amongst millions, and maybe the Earth would be better without humans. World population growth is clearly not sustainable, and pressure on natural environments and biodiversity is increasing. Mankind faces many moral dilemmas in his relationship with nature and exploitation of resources, and the standard stopgap response seems to be that we destroy a lot and maintain a small part as a 'reserve' or 'park'.

For mankind to live in tune with nature in a sustainable way, our modern civilisation would have to revert to a pre-industrial society. Indeed, some would say we need to retreat to the hunter–gatherer lifestyle. The only way I can see that this might happen is through the rapid spread of a worldwide disease that decimates the human population by more than 95%, and results in the collapse of political institutions, transport, power and modern electronic communications.

On a geological timescale change is inevitable. The regularity of the Earth's orbit around the sun ensures that the climate alters with periodicities ranging from about 20,000 to 400,000 years. It would be surprising if another ice age did not return and cover Scotland with ice within the next 20,000 years. If mankind is still as abundant as today, it is not difficult to imagine the political problems that would occur as people all try to migrate to more favourable areas: war would seem inevitable, probably using religion as an excuse.

Another scenario is the eruption of a 'giant igneous province' when magma is discharged in quantities far greater than anything that has been experienced during the existence of mankind, although such events are evident in the geological record. The resulting rapid climatic changes would be worldwide and rapid, providing a terrible obstacle to the survival of civilisation. I would not wish to go forward in time by 10,000 years, but I would like to see the state of Earth in ten million years' time. My guess would be that there would either be no sign of mankind, or there would be a highly developed society of hominids evolved from mankind.

Speculation and imagination of the future will always be popular in science, art and literature, but there are few 'facts' to rely on and only the opinions of computed models. On the other hand the few short journeys to the past that have been described in this book are supported by geological evidence and the fossils of the animals that lived at the time. As our knowledge of the past increases, the inescapable conclusion is that the world and its biota are constantly changing, and will always do so. Some animal and plant groups have remained remarkably constant, or conservative, through hundreds of millions of years, while others have gone through phases of explosive rapid evolution, only to be wiped from the face of the Earth. Is mankind to be in the former or latter category? I leave you to decide for yourself.

Bibliography

Palaeontology (selected texts to university level)

Benton, M. J. (2003) *Vertebrate Palaeontology* (3rd edition), Oxford: Blackwell Publishing

Clarkson, E. N. K. (1998), *Invertebrate Palaeontology and Evolution* (4th edition), London: Blackwell Science

Cleal, C. J. and Thomas, B. A. (1999), *Plant Fossils*, Woodbridge: The Boydell Press

Fortey, R. (2002), *Fossils: The Key to the Past,* London: Natural History Museum

Kenrick, P. and Davis, P. (2004), *Fossil Plants,* London: Natural History Museum

Long, J. A. (1995), *The Rise of Fishes, 500 Million Years of Evolution,* Sydney: UNSW Press

Milsom, S. and Rigby, S. (2003), *Fossils at a Glance*, Blackwell: Oxford

Monks, N. and Palmer, P (2002), *Ammonites,* London: Natural History Museum

Taylor, P. D. and Lewis, D. N. (2005), *Fossil Invertebrates,* London: Natural History Museum

Willis, K. J. and McElwain, J. C. (2002), *The Evolution of Plants,* Oxford: Oxford University Press

Palaeontology (guidebooks for fossil identification)

Murray, J. W. (ed.) (1985), *Atlas of Invertebrate Macrofossils,* London: Longman Group

Natural History Museum (1975), *British Palaeozoic Fossils* (4th edition), London: Natural History Museum

Natural History Museum (1983), *British Mesozoic Fossils* (6th edition), London: Natural History Museum

Walker, C. and Ward, D. (1992), *Fossils. Eyewitness Handbook,* London: Dorling Kindersley

Palaeontology website

Trewin, N. H., Fayers, S. R. and Anderson, L. I. *Early Terrestrial Ecosystems: The Rhynie Chert,* www.abdn.ac.uk/rhynie (documents the geology and palaeontology of the Rhynie and Windyfield cherts)

General Scottish Geology

Clarkson, E. N. K. and Upton, B. G. J. (2006), *Edinburgh Rock: The Geology of Lothian,* Edinburgh: Dunedin Academic Press

Gillen, C. (2003), *Geology and Landscapes of Scotland,* Harpenden: Terra Publishing

McKirdy, A. (series ed.; various dates), *Landscape Fashioned by Geology*, Battleby: Scottish Natural Heritage (relevant to this book are the titles on *Scotland; Edinburgh and West Lothian; Skye; East Lothian and the Borders;* and *Fife and Tayside*)

McKirdy, A., Gordon, J. and Crofts, R. (2007), *Land of Mountain and Flood, The Geology and Landforms of Scotland,* Edinburgh: Birlinn

Trewin, N. H. (ed.) (2002), *The Geology of Scotland* (4th edition), London: Geological Society

Upton, B. G. J. (2004), *Volcanoes and the Making of Scotland*, Edinburgh: Dunedin Academic Press

Geological Excursion Guides (including excursions relevant to this volume)

Bell, B. R. and Harris, J. W. (1986), *An Excursion Guide to the Geology of the Isle of Skye*, Glasgow: Geological Society of Glasgow

McAdam, A. D. and Clarkson, E. N. K. (1986), *Lothian Geology, An Excursion Guide.* Edinburgh: Edinburgh Geological Society and Scottish Academic Press

MacGregor, A. R. (1973), *Fife and Angus Geology*, Edinburgh: Scottish Academic Press

Trewin, N. H. and Hurst, A. (1993), *Excursion Guide to the Geology of East Sutherland and Caithness*, Edinburgh: Geological Society of Aberdeen and Scottish Academic Press

Trewin, N. H., Kneller, B. C. and Gillen, C. (1987), *Excursion Guide to the Geology of the Aberdeen Area*, Edinburgh: Geological Society of Aberdeen and Scottish Academic Press

British Geological Survey (various dates), *British Regional Geological Guides* to the areas of Scotland contain much detailed local information

Gazetteer

Entries in **bold** denote illustrations. National Grid references are given for selected geological localities visited on the excursions in this book.